ROUTLEDGE LIBRARY EDITIONS:
HUMAN GEOGRAPHY

Volume 16

NATION, STATE, AND TERRITORY

NATION, STATE, AND TERRITORY
A Political Geography

ROY E.H. MELLOR

Routledge
Taylor & Francis Group

LONDON AND NEW YORK

First published in 1989 by Routledge

This edition first published in 2016
by Routledge
2 Park Square, Milton Park, Abingdon, Oxon OX14 4RN

and by Routledge
711 Third Avenue, New York, NY 10017

Routledge is an imprint of the Taylor & Francis Group, an informa business

British Library Cataloguing in Publication Data
A catalogue record for this book is available from the British Library

ISBN: 978-1-138-95340-6 (Set)
ISBN: 978-1-315-65887-2 (Set) (ebk)
ISBN: 978-1-138-99904-6 (Volume 16) (hbk)
ISBN: 978-1-315-65845-2 (Volume 16) (ebk)

Publisher's Note
The publisher has gone to great lengths to ensure the quality of this reprint but points out that some imperfections in the original copies may be apparent.

Disclaimer
The publisher has made every effort to trace copyright holders and would welcome correspondence from those they have been unable to trace.

NATION, STATE, AND TERRITORY

A Political Geography

ROY E.H. MELLOR

London and New York

First published 1989
by Routledge
11 New Fetter Lane, London EC4P 4EE

Simultaneously published in the USA and Canada
by Routledge
a division of Routledge, Chapman and Hall Inc.
29 West 35th Street, New York, NY 10001

© 1989 Roy E.H. Mellor

Reprinted 1991

Phototypeset in 10pt Times by
Mews Photosetting, Beckenham, Kent

Printed in Great Britain by
Antony Rowe Ltd, Chippenham, Wiltshire

British Library Cataloguing in Publication Data
Mellor, Roy E.H. (Roy Egetron Henderson)
    Nation, State and territory : a political
    geography.
    1. Political geography
    I. Title
    320.1′2

    ISBN 0-415-02287-8

Library of Congress Cataloging-in-Publication Data
Mellor, Roy E.H.
    Nation, state, and territory : a political geography / Roy E.H.
    Mellor.
        p.   cm.
    Bibliography: p
    Includes index.
    ISBN 0-415-02287-8
    1. Geography, Political.   I. Title.
    JC319.M56   1989   88-23900
    320.1′2—dc19   CIP

Contents

Contents

Figures

Introduction

A political dimension is seldom absent from human geography, for political ideologies and policies influence most decisions on social and economic issues. Whether we consider settlement or land use, transport or industrial location, the geographical pattern, even elements of the visual landscape, are likely to be affected by political perceptions and decisions. To a considerable measure the spatial pattern of mankind is moulded by the interactive relationships within a tripartite association between nation, state, and territory, the core of the aspect of the discipline we term 'political geography'.

The elements of this association are important to us all, for none of us can escape them, and exert a powerful influence over our everyday life. Every one of us is a member of an ethnic and a political nation; each of us must live within the rules set in law by the government of the state in which we live; and every human being has a natural feeling of territoriality at whatever scale we define that.

The spatial expression of the combination of nation, state, and territory is the chequered quilt of the political map. The pattern of the quilt as we see it at any point in time is the product of historical evolution, to which change has come as the relationships and balance between the three elements have shifted. Such change is ongoing and even today we need to look neither far nor deeply to witness its progression. In the political map we see people's sense of community as the different levels build into the ultimate expression of the nation. It also reflects how nations choose to organize their political life by forming states to secure their welfare and independence, whose governments interpret society's values as political ideologies, turning them into policies that have a deep imprint on the landscape. The real geographical dimension in the tripartite association is given by the national territory, reflecting a fundamental human response of attachment to a homeland. Napoleon expressed this rather deterministically but not without truth in his dictum that the policy of a state is decided by its geography.

Behind most political geographical problems there is usually a significant historical dimension and this study makes no apology for giving it some prominence. Traditions and attitudes, perceptions and values that form the substance of national identity and aspirations have strong historical roots built into the nation's educational system, so consequently colouring perceptions and reactions to new situations and the options available for their solution. The relationships between geography and history are deep, complex and ever present, though often difficult to interpret: the Russian historian Vernadsky believed 'the logic of geography lies at the root of all history', but we may suggest that at any point in time the geographical pattern of mankind bears a strong imprint of the not uncommonly fortuitous events that chart the tortuous course of historical experience.

This study arises from teaching political geography for many years and is coloured by my long-standing regional interest in Europe and the Soviet Union, from where consequently many of the illustrations are drawn. My interest in political geography was first stimulated by the late Professor Walter Fitzgerald, and subsequently developed through encouragement from a former Geographer to the US State Department, the late G. Etzel Pearcy.

My thanks go to my colleague, Dr E.A. Smith, CBE, who read and commented on the manuscript, giving up valuable time from a busy schedule. Of course, remaining shortcomings and the views expressed are entirely mine. Mrs Jane Calder meticulously typed a most untidy manuscript, while Mr Lawrence McLean and his team of cartographers turned messy sketches into respectable illustrations. Throughout, my wife coped with many time-consuming but unexciting editorial tasks and kept me ticking along with endless cups of tea. At the publishers my flagging efforts were given a new lease of life by gentle enquiries and encouragement by Mr Peter Sowden, while Miss Anita Roy, Mrs Margaret Baker and Mr Gary Davis did noble editorial work. To all, a deep and heartfelt thank you.

1

The nation

The pattern of the political map is moulded by people, individuals acting collectively through a hierarchy of community to shape and direct the political process and government. In democratic regimes people have an opportunity to voice their opinions on how political matters should be managed, but in autocratic or totalitarian regimes it is only those who enjoy the inherited privilege of belonging to the ruling elite (or who have otherwise worked their way into it) who can exercise influence in these matters. For the political geographer the most significant level in the hierarchy of community is the nation — the highest level of identity. The modern concept of the nation crystallized in philosophical debates in the eighteenth century and grew powerfully in expression and coherence in the nineteenth and early twentieth centuries. Though social scientists have distinguished between 'ethnic' (or 'cultural') nations and 'political' nations, a quite useful distinction, it is perhaps better in political geography to limit the term 'nation' only to groups that have achieved political recognition and have created their own sovereign state. Other groups that have not achieved that status may best be termed ethnic groups or 'ethnies', even though many aspire to nationhood in the political sense.

In modern times the nation-state, representing the sovereign-state apparatus and legal nationality of the dominant ethnic group in its territory, has superseded other forms of sovereign state such as that organized around a dynasty. It is important to recognize that every individual has both an ethnic nationality and a legal nationality, which usually but not necessarily coincide. Peoples aspiring to political recognition as nations are in a sense 'sub-nations', a term not infrequently met, and often enjoy some measure of autonomy within another people's nation-state, although sadly some receive little separate recognition or may even be repressed and persecuted.

When aspiring to nationhood, ethnic groups with a well-defined settlement area are in a stronger position than those scattered in penny packets across another's homeland. The coherent settlement area of the Basques in northern Spain is an important weapon in their struggle for autonomy, whereas the Jews, after their diaspora in Roman times, faced the dilemma of being scattered in small groups mostly in towns across the Empire. The pressure for political identity and autonomy by aspirant nations has underlain much instability in the political map. It was a major factor in the disintegration of the dynastic Habsburg Empire and the recasting of the map of East Central Europe after 1919. But instability may also be generated when a nation becomes expansionist, seeking new settlement areas or the domination of other peoples, as demonstrated by the Japanese in East Asia earlier this century.

Though common bonds may long have laid dormant, formation of a national identity requires that innate sentiments are encouraged and codified to coalesce into nationhood. The leadership may come from an elite or aristocracy or even from the broad ranks of society, but certainly the intelligentsia will play a key part in codifying the national language, generating a national literature, gathering music and recording folklore, legend, and history. It is usually possible to identify a focal area or core where the national movement is stimulated and from which national aspirations are disseminated to recruit new members. Later, this core area often acts to create centripetal forces to draw the nation together, for within the perceived national territory, there are sometimes groups of suspect loyalty whose influence generates destabilizing centrifugal tendencies, so that the core must act to counter them.

DEFINING THE NATION

A nation may be described simply as comprising people sharing the same historical experience, a high level of cultural and linguistic unity, and living in a territory they perceive as their homeland by right. There is a further but less readily definable dimension in the mental perception of the members of the group of their common heritage and 'togetherness'. This spiritual or metaphysical dimension was well summed up in Kipling's poem 'The Stranger'. In nationhood, people with common traits and aspirations achieve recognition of their communal identity and their right to do things their own way, usually termed their 'way of life'. The emergence of national identity has usually been a lengthy process around which has been built an elaborate symbolism, using events, personalities, and even places from historical experience, often given an almost mythical quality as the national 'iconography'. The interpretation of this iconography is passed from generation to generation through the educational system which inculcates the mores — the values and customs — of the nation into its members, seeking to stimulate patriotism and national pride. Because man has a strong feeling of territoriality in his make-up, the territorial dimension is a powerful element in national iconography, reflected in such mental images as *Blighty* for the British, *die Heimat* for Germans, or the Russians' *Matka Rossiya*.

Through the strength of attachment to what it believes to be its 'national territory', the nation often comes into conflict with neighbours, who may also covet part of it. The geographical analysis of territorial disputes forms an important part of political geography, but the geographical interest is even wider, since the nation's 'way of life', its social and economic expression, makes a deep impress on the landscape. This latter impress does, of course, inherit much from the political ideology under which the nation lives and, while this is often a free choice of the majority of the members, it may be imposed from outside or by a powerful ruling elite that gains much of its strength from external support.

The political expression of the nation's aspirations we may term 'nationalism'. Nationalism uses the elements and symbolism of national identity to express the objectives of the nation's political grand strategy. It builds on the nation's hope and fears, arousing popular support by appeals to national pride and to patriotism, not uncommonly couched in a promise of a new golden age, often resurrecting a popular view of some historical or even near mythical episode in the nation's development. To attain and sustain national identity and win recognition of nationhood, a political strategy is inescapable, and consequently every nation has its nationalism. When that nationalism contains aggressive or expansionist, even racist traits, its jingoism can pose a scenario for conflict, often emotively presented as a 'struggle for national liberation'.

Seton Watson (1977) noted that a 'scientific' definition of nation had never been achieved and probably never will be. Each nation is unique in the way its elements are combined, let alone how successfully they have jelled into a durable identity. The most significant elements in nation-building are language, religion, and historical experience, but there are also more unquantifiable ones such as custom and usage or the sense of togetherness. It is usually difficult enough to define what holds one's own nation together, let alone any other. If we take elements like language and religion, even custom and usage, it is difficult for an outsider to understand why, say, Bavarians and Austrians do not feel themselves close enough to merge into a common identity — is it historical experience or some other element that separates them? Historical experience might certainly account for their separate identities, but there also seem to be far less tangible qualities that motivate their distinctiveness. It is equally difficult to appreciate why in some instances (the Swiss, for example) strength of common historical experience and a common consensus of aspirations have been sufficient to weld into nationhood groups without a common linguistic or cultural background. No particular combination of elements seems to make inevitable the generation of a national feeling among a group, nor is such sentiment precluded simply because a particular trait is absent.

LANGUAGE

We all learn a 'mother tongue' as children and some of us later become fluent in other languages. Language is central to any group as a means of communicating ideas and instructions, though each language by its nature has an important influence on how ideas, concepts, values, and imagery are expressed. Language, culture, and perception are intimately intertwined. It is thus not surprising that language has been a key element contributing to a sense of national identity. Superficially, at least, it is one of the easiest indicators to quantify ethnic groups: census-takers have commonly defined 'ethnic affinity' on the basis of 'mother tongue' or other attribute of language.

In Europe, language as an identifier of national groups came with the shift to the vernacular in law and administration from the four-teenth century. The development of a standard written form received a considerable stimulus with the introduction of printing. A further stimulus came in the eighteenth century as interest awoke in folklore, dialects, and oral traditions, which generated additional codification and classification of languages. Compulsory education and rising literacy spreading the readership of newspapers and books during the nineteenth century, as well as the influence of military conscription, added still more encouragement to the use of standard national languages.

The evolution of a standard national language often provides clues to the geographical pattern of national emergence, indicating the core area from which a national movement disseminated. Standard written and spoken English is marked by the rise of the East Midlands as a focus of population and economic strength from the fourteenth century, especially with the increasing concentration of political power. It coincided with the use of English instead of Latin by the judiciary and the influence of London was enhanced as it grew as the centre of printing and publishing in the fifteenth century. The late and incomplete submergence of Scotland under the English Crown is still reflected in differences of usage and vocabulary, for standard English did not usurp the older Scots tongue until the seventeenth century.

The process was slower and less complete within the German lands of the Holy Roman Empire, where Latin remained the medium of law and administration, but the various princely chancelleries increasingly used the local German speech in deeds and documents, and by the late fifteenth century, the Emperor Maximilian had managed to establish a reasonably standardized usage. The Imperial Chancellery in Prague had given the initial leadership, but with the decline of the Luxemburg dynasty, however, further development passed to the Wettin Chancellery in Saxony. This usage was strengthened when Luther used it for his translation of the Bible and it became the preferred style of book publishers. By the mid-eighteenth century, long before national unity was achieved, the German states had a standard written literary language and this 'paper' language became the spoken medium, a development hastened by political unity after 1871. Local dialects are still widely used, but in the larger towns the spoken language lies between dialect and the norm taught in schools. Regional associations carefully nurture dialect, reflecting the strong regional element in German life. Differences remaining between north and south are perhaps reflections of the lack of an early focal point in the growth of national identity, while the distinctiveness between Berlin and Vienna in speech and usage accentuated the feeling of separateness between the Hohenzollern and Habsburg empires.

The Low Countries saw their Platt German dialects, in the beginning closely related to Low Saxon and Rhine Frankish, emerge as a distinct literary language. The common origin as a West Germanic tongue is inherent in the name *Dutch* (the same as *Deutsch* (German)) though *Nederlandsch* came into use in the seventeenth century, becoming the official name in 1815. The modern standard language is based on the Holland dialect of Amsterdam, where political and economic power lay in the formative seventeenth century. The separation of Dutch from German arose from the active literary life in the local dialects from the middle ages onwards. The concentration of mercantile towns, compared with the more sparsely settled areas of Low German speech in the broader geographical framework, created a healthy milieu for the preservation of local usage, while the upper and merchant classes, strongly influenced by cultural streams from the south, found little affinity with the 'High' German tradition. Furthermore, the struggle against Spanish domination made language a significant element in national identity before it came to fill that role in other German lands. Acceptance of Calvinism tended to weaken links with the Catholic Rhineland and Lutheran North Germany. Some divergences remain between the resolutely Catholic Brabant and Flemish lands and the remaining provinces.

Although language has not been a prime factor in the emergence of national identity in Switzerland, one might have expected the German cantons in their long struggle with the Holy Roman Empire and the Habsburg to have developed a linguistic identity on the model of the Netherlands. Though the distinctive Alemannic spoken dialects are far removed from the written idiom of Saxon German used by Luther, it has been suggested that acceptance of this latter form as the official language, so patently different from everyday speech, arose because Zwingli chose to use it to dispute articles of faith with the Lutherans in their own written form. It has also been suggested that the local dialects were too jealously regarded to allow evolution of a synthetic written and spoken Alemannic standard form.

Language played a key role in the early emergence of a close-knit national identity in France. Once again, the ascendancy of one region, the Ile de France, has left its impress, with the early romantic literature in the Midi, using Provençal, eclipsed by the written and spoken dialect of the politically vigorous north. The early sixteenth century marked a drive for unification as fears of outside pressures grew, reflected in the royal decree of 1539 that only the northern form of French should be used in provincial chancelleries. By the seventeenth century the written language of the Parisian bourgeoisie had come to be accepted as the norm, though its use in speech took longer to establish. The local dialects, the *patois*, were usurped by a standard spoken language as the centralizing tendencies initiated by the Revolution spread, reinforced later by compulsory military service and universal schooling. The distinctive regional speech of the early Romance period has been almost obliterated, though provincial ways remain of pronouncing the standard language, but the nature and content of that language and its usage have been jealously guarded since 1635 through the Académie française. A significant element in the iconography of

8

French national identity is a belief in the superiority of their language ('what is not clear is not French').

The long-standing identity of the Chinese nation has arisen from the early distinctiveness of their unique civilization, setting them clearly apart from the peoples around them, whom they regarded as grossly inferior. The development of an ideographic script (representing not sounds but concepts) has meant that, however divergent the local spoken vernacular, the written symbols are intelligible throughout the vast territory (Fig. 1.1). Though north and south cannot understand each other in speech, they can do so in writing. For the small educated class, the intelligentsia of the administration, this was a valuable tool, but the complexity of the script hindered the spread of literacy. For an industrializing country, this has been a serious drawback, so the Chinese are now faced with the monumental task of developing a common spoken language, primarily based on the northern dialects, notably that of Peking (the seat of political power).

Figure 1.1 Languages of China.

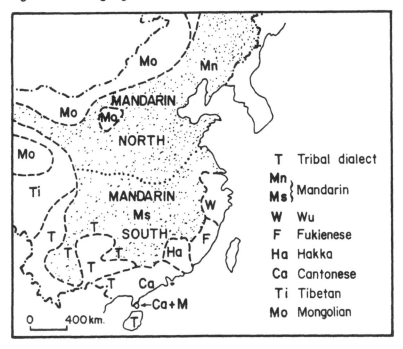

T	Tribal dialect
Mn	
Ms	Mandarin
W	Wu
F	Fukienese
Ha	Hakka
Ca	Cantonese
Ti	Tibetan
Mo	Mongolian

Adapted from *Admiralty Handbook, China*, vol. I, London 1944

At the same time, to ease dissemination of literacy, they are attempting to replace the ideographic script by an alphabetic romanization of the written language, though because Chinese is strongly tonal in character, this is a daunting task.

Attainment of independence can present problems of finding a suitable official language to replace that used by the displaced colonial authorities. The Irish Free State in 1922 set out to create a separate Irish identity after centuries of English rule and felt language was an important medium in the task. During the long English domination the Irish Celtic tongue had become a minority language, losing its literary tradition and retreating to the remoter parts. A resurrected and revivified Irish language was introduced into the education system and a vigorous attempt made to force its use through government agencies. Nevertheless, after sixty years, the population remains overwhelmingly English-speaking, perhaps reflecting the strength of commercial and other ties with the rest of the English-speaking world.

Efforts to create a sense of national unity, if not a single identity, in India since Independence have been bedevilled by the language issue. Despite the great number of languages spoken, six of them account for two-thirds of the population, while well over 90 per cent of the people are encompassed by the fourteen 'languages of India' defined in the Constitution. Such linguistic diversity has been seen as posing a primary threat to national cohesion, especially as no one language is spoken by sufficient people to give it undisputed superiority. Hindi or Urdu or other northern languages are, however, not readily acceptable to the quarter of the population using earlier Dravidian languages in the south (Fig. 1.2). Although the Constitution had recognised Hindi and thirteen other regional languages (four from the Dravidian family), there remained a need for one language to be a nationally recognized medium to replace the imperial *lingua franca*, English. Policy strongly favoured Hindi (used by just under a third of the population) in the Delhi dialect, support for which as the 'official language of the union' had been given by Gandhi. Nevertheless sectarian linguistic discontent caused civil disobedience, so the Delhi government, against its wishes, was forced to placate the regional interest by drawing internal territorial-administrative boundaries along linguistic lines. The government view has been that this has generated centrifugal forces when the need has been for centripetal ones in the interest of national economic and social welfare. English has been retained as an administrative language long after the original intention and Hindi's role has been scaled down in the south. In the school system local languages are used, but Hindi is required at secondary level and above, while English is optional for government positions.

Figure 1.2 Languages of India.

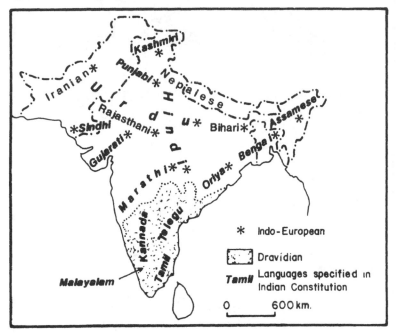

Compiled from various sources

Switzerland is usually cited as an example that multilingual nations can be created, and there is no doubt that the Swiss are keenly aware of their common identity. Nevertheless Switzerland displays some special aspects, such as a federal state of a strikingly decentralized kind, allowing the cantons a substantial degree of freedom to manage their own affairs. It holds together well because the historical experience of the people has produced a strong conviction of common action to preserve their way of life against outside pressures. At the same time there has been a sensible approach to language, with care taken to find compromises over linquistic issues, especially important during major crises like the two world wars, when relations between the striking predominance of German speakers and the Francophones required particularly careful management. The official recognition in 1938 of Romansch simply confirmed a *de facto* situation, with the

11

Figure 1.3 Language and religion of Switzerland

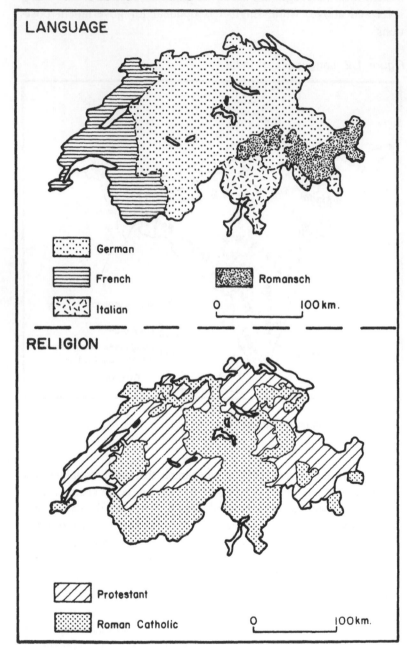

Adapted from *Schweizerischer Mittelschulatlas*, Zurich 1975

many different dialects brought together in an agreed written language (Fig. 1.3). One important factor has been that Swiss of whatever mother tongue have regarded the other languages as equals, with a remarkable level of multilingualism.

National coherence in Norway was faced by distinct and not always easily mutually intelligible dialects, contributing to powerful local lobbies in national life. These grouped into two main forms — *Bokmål* or *Riksmål*, used notably in the bigger towns and in the southeast, and *Nynorsk* or *Landsmål*, spoken especially in the west. Last century language became a focus of a desire to strengthen Norwegian cultural traditions, infiltrated by the closely related but quite distinct Danish during the union of the Norwegian and Danish crowns from 1380 to 1814. *Nynorsk* was a particular product of this movement. When Norway became fully independent after the union with Sweden (1814–1905), *Bokmål* and *Nynorsk* was given equality (1907) and through subsequent reforms have tended to draw together. A true national language has not, however, been created, with the reforms resented by *Bokmål* speakers, claiming they favoured *Nynorsk*. A mutually intelligible written and spoken form has emerged for use in the press, broadcasting, television, and education, though the vernaculars remain. Currently, about one-fifth of the people use *Nynorsk*, the remaining four-fifths *Bokmål*, especially in the towns.

In Belgium efforts to generate a true Belgian national identity have been thwarted by implacable linguistic animosities. The division between French Walloon speakers and Flemish speakers has been reasonably even and there has been a remarkable unanimity in religion, unlike the violent religious confrontations that for so long spent Swiss energies (Fig. 1.4). Despite the saying 'Flamand, Wallon ne sont que les prénoms — Belge est notre nom de famille', an insurmountable difficulty has been that the Walloon element (particularly the Francophone intelligentsia) has always regarded Flemish as an inferior language and has sought to assert the use of French as exclusively as possible in public life, though the early advantage of the Francophones has been eroded as economic strength shifted into Flemish districts. Francophone hostility and the rise of Flemish economic muscle acted significantly in creating a feeling of Flemish nationhood and the need to create a Flemish cultural and economic sphere. The Francophone population consistently argued for a 'unitary' Belgium, believing their linguistic superiority would still give them the advantage, particularly through growing French influence in Brussels itself. How the interests of the Francophones, the Flemings, and the small

13

German minority can be satisfied and a strong national feeling generated remains unsolved, despite major constitutional amendments, particularly as party allegiance is divided along linguistic loyalties.

Figure 1.4 Language divisions in Belgium.

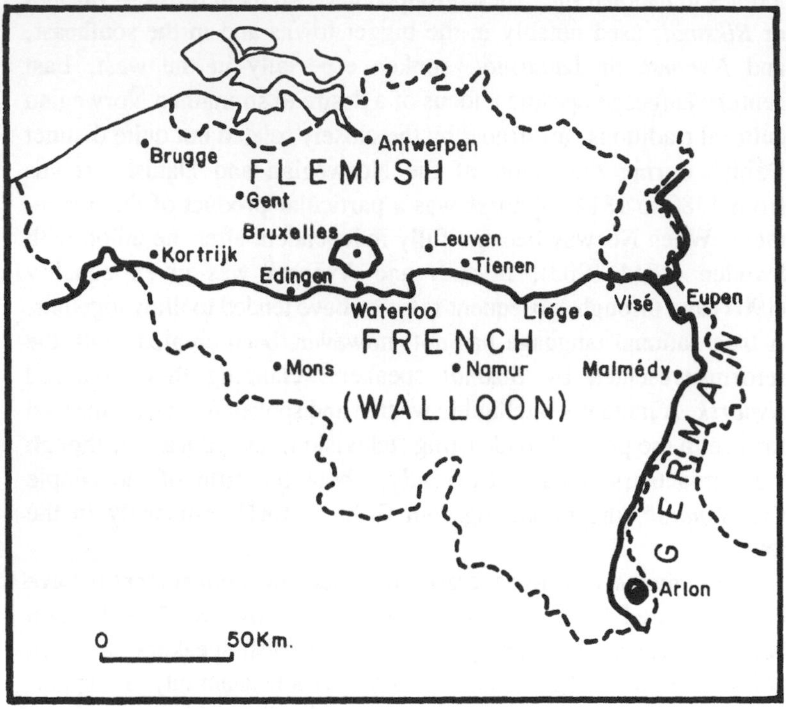

Compiled from various sources

Language has commonly been used as an indicator of ethnic membership in national censuses, though the published results have often been 'manipulated' for political ends. There is commonly difficulty in distinguishing 'languages' from 'dialects', on which even experts cannot always agree. The Prussian censuses before 1914 claimed both Kassub and Masurian as distinct languages (whereas the Poles maintained they were merely dialects of Polish), so reducing the size of the Polish population in their eastern provinces. The Poles likewise in their interwar censuses attempted to establish the language status of Ruthenian as a pretext to claims over territory for which no other justification existed. Habsburg censuses had likewise

14

swelled the number of Germans in their provinces by counting as such Yiddish speakers, while the French reduced the number of German speakers in Alsace by distinguishing between High German and the local Alsatian dialect. Bias may arise from the way the census question is asked, how it is answered, or from interpretation. To ask for 'mother tongue' (the language first used in the home) favours minorities, whether in process of assimilation or not, while asking for the 'usual' or 'customary' language (that used in everyday affairs) tends to emphasize the official or dominant language. In the end all depends on a truthful response — in the Balkans enumerators are known to have pressurized individuals to give the 'right answer'; elsewhere many people have admitted to giving the 'right answer' to help their careers. Just as many middle- and upper-class Welsh after the Tudor conquest accepted the English language to advance their social and economic welfare, so many Flemings in nineteenth-century Belgium took a veneer of French language and culture for the same reason, while numbers of Poles, Czechs, Magyars, and others cloaked themselves as Germans in the Habsburg Empire.

States have applied forceful integration of minorities by enforcing the use of the 'national' language. Before 1919 the Hungarians were notorious for their magyarization programme, while the Prussians pursued germanization in their eastern territories, with Polish retreating in Silesia, notably when industrial development took off in the late nineteenth century. In Tsarist Russia every effort was made to weaken languages other than Great Russian, even prohibiting publication in Ukrainian and Byelorussian. Although in the Soviet Union there is now a quite liberal language policy, anybody hoping for career advancement finds it imperative to have a fluent written and spoken command of Great Russian. There is, of course, a tendency among bilingual populations to be ambivalent and opportunist towards nationality. Political pressures on minority language groups can change their geographical distribution. From the late eighteenth century Paris sought to strengthen its hold over the national territory by imposing a uniform educational system in the French language which, along with social and economic factors, has resulted in a decline in Flemish near Dunkirk, a retreat into the less accessible west of the Armorican Peninsula by Breton speech, and the contraction in the south of both Basque and Catalan. Its effect in the more fluctuating political fortunes of the northeast forced only a modest retreat of the French-German linguistic frontier in Alsace and Lorraine. The position of German in Alsace has been strengthened in the last quarter century through the influence of television programmes from across the

German border and by closer economic ties with Germany.

Political difficulties may arise from even a liberal language policy, as in Canada, where English and French have equal status and the federal structure allows provinces management of many of their own affairs. Spurred by the anglophobe de Gaulle to reaffirm their Frenchness and fearing the relentless pressure of the English language media, the French Canadians voted in a Quebec provincial government in 1976 to impose a new policy favouring French and containing English (the so-called *Loi 101*). This has added fuel to fears of Quebecois separatism, but it is hard to see how it can benefit the province in its minority language position in the English-dominated commercial and political life of North America.

Some countries have tried to sweep their language problems out of the political arena by make-believe that they no longer exist. In Belgium since 1960, for example, collection and publication of linguistic statistics has been forbidden by law. This seems to give rumour an open door: already quite exceptional percentages are being quoted for the growth of Flemish speakers without any sound evidence.

Language is a powerful political weapon in an age of mass communication and a national language is seen as a key factor in creating national coherence. The choice is not always a ready one and often gives rise to internal strains. Minority peoples seeking political recognition centre much effort around their language, especially where they seem threatened by major and often more virile languages. Claims for equality of minor languages with major and national languages have become increasingly common as political issues.

RELIGION

Religion (especially through sectarian conflict) has played a significant role in the emergence of many nations. This was certainly a most crucial element in the past, though in more recent times its role appears to have declined. The role of religion has not only been in forming values and aspirations but also in moulding group identity through frictions between different beliefs. Frictions within the Christian ethos in Europe have been a powerful contributor to the groupings under national banners. The first major divide in the Christian tradition came with the Great Schism of 1054 that left a deep impress on later national development. In the west nations began their slow evolution within the orbit of the Roman Church, with its universalist Latin tradition and alphabet, whereas in the east national evolution was guided by the Byzantine Orthodox tradition and use of the Cyrillic script. Five centuries later the split of the Protestants from the Church of Rome made a further indelible mark on the map as tensions between the two denominations strengthened identities as one group sought to maintain itself against the other. In the formation of kingdoms around which several nations were to crystallize, the Church of Rome had played an important role. The Pope sought to maintain the secular tradition of the Empire by creating the Holy Roman Emperor. Between them they elevated rulers to kingships and invested them with an element of divine right that ensured they would be held in absolute respect. Among those so elevated were the rulers of Hungary, Poland, Bohemia, and Sicily, all territories where an early form of national identity was to emerge, if not to mature.

Religion has played both an internal and external role in the development of nationhood. Polish identity began to emerge after acceptance of Christianity by Prince Mieszko in AD 966, and further consolidation came with medieval crusades to spread Roman faith among the pagans of the Lithuanian 'wilderness' and the Orthodox Christians of the Russian marches. The final triumph of this evangelism was acceptance of the faith by the Lithuanians, when their ruler in 1386 saw such an act could lead him to the Polish crown. Polish nationhood subsequently retained a strong element of bearing a banner for the Church of Rome on the frontier with the Russian Orthodox Church and against the Protestant ethic strongly entrenched in Prussia. The struggle against the Albigensian heresy in southern France in the thirteenth century strengthened the hold of the French crown and helped to consolidate the identity of the French nation. The bitter turmoil of the Reformation in the sixteenth century was an important phase in the emergence of national identity in Europe. Much of the ultimate territorial extent of Protestantism was determined by the decision of rulers (*cuius regio, eius religio*). Protestantism in mainland Britain and its struggle against Catholic Spain in Tudor times were a powerful influence on the jelling of British nationhood. Even so, we may see the internal divisions of Protestantism, notably Scottish Calvinism's suspicions of the episcopalian churches, as a factor in the preservation of a distinctive Scottish identity. An inflexible British administration plus a Protestant presence, said to encourage discrimination against the Roman Catholic majority, were elements in fermenting Irish national identity, culminating in the forma tion of the Irish Free State in 1921. The Protestant-Roman Catholic confrontation, bitter with bigotry almost unbelievable in this day, remains a destabilizing force by being made a cloak for political conflict in Ulster.

The Orthodox Church, unlike Rome, was not universalist, so the ethnic patriarchs provided a focal point for emergent national feelings. They proved their value during the long Ottoman period, when a tolerant Islam allowed them the chance to maintain the identity of their flocks. The Orthodox Church made little effort to win new converts, content to survive through containment within its own ethnic association as a strong national identity emerged among such peoples as Serbs, Greeks, and Russians, focused upon their patriarchs. It is interesting to conjecture how different the internal problems of the Soviet Union would now be, had the Orthodox Church sought vigorously to win converts from the Muslims of Central Asia and the Volga.

Irreconcilable hatred between Muslims and Hindus forced the partition of the Indian subcontinent at Independence. Religious ties were, however, not enough to create a national identity between the contrasting cultural and linguistic patterns of East and West Pakistan, with the former breaking away as Bangladesh. Within Islam, the emergence of a new militancy among the Shi'ites poses a destabilizing influence, often in strained sectarian relations with other interpretations of the faith. This has been well seen in the bitter clash with Maronite Christians and even with rival Muslim factions in the Lebanon, where national identity has been torn apart, and also in the Iran–Iraq clash. Islamic precepts have in the last thirty years taken a powerful new role in the iconography of nationhood in several Muslim countries, such as Libya and Iran. In parts of Africa competition between Islam and Christianity for converts may well have an important influence on the way nationhood develops among some of the recently independent former colonial territories.

Where toleration and acceptance exist, powerful national identities have developed among people not all of the same religious conviction. In West Germany, the Roman Catholic–Protestant dichotomy has now little influence on the feeling of nationhood, though it still plays a role in regional politics and may also extend to national level. The same is true in Britain and in the United States of America. In Poland increased popular support for the Roman Catholic church (to which most of the population belongs) has been an act of defiance against communist policies. In Yugoslavia regional tensions between north and south arise more from contrasts in economic conditions and living standards than from religious strains between the Roman Catholic north and the Orthodox south, though Yugoslavs are fully aware of the need for a strong overall national solidarity.

CUSTOM AND USAGE IN THE NATION'S ICONOGRAPHY

Custom and usage, important in daily life and interaction in the community, are a significant element in the many complex factors that weld communities into nations, but are among the least readily quantifiable parameters. The eighteenth-century interest in folk tradition, in costume, in music, in language stimulated national feelings in Europe, and we have come to accept 'national' schools in painting, in music, in literature, architecture, and other skills. Much early study of these elements was done by scholars whose findings were frequently romanticized and 'improved', to be built into the educational process through which everybody passes, so that from childhood members of the nation are inculcated with its interpretation of national history and values, instilling pride and patriotism. These symbols comprise what Gottmann has described as 'national iconography', powerfully linked to the sacrosanct national territory, capitalizing on the urge to define as clearly as possible exact limits to territorial control exercised at every level of the hierarchy of community.

These elements build together into the 'national way of life', combining with economic patterns, but there is often reference to a tantalizingly elusive quality, usually called 'national character'. Many social scientists reject the existence of this quality, but people who have lived among another nation for a long time do not so readily reject it. It is very much a group characteristic and doubtless we know from our own experience that we show different sides to our character and mental processes as individuals rather than as members of a group. Even during a short holiday abroad we have probably felt a greater sense of strangeness towards the locals as a group than towards them as individuals. We see 'national character' as much in our own group as in others, especially when we compare our 'virtues' with other peoples.

Our mental image of other peoples is so often revealed in caricature in films or literature or even in the way others are presented in news items. To describe national character sympathetically, sensitively, and reasonably objectively requires great skill, great experience, and great compassion, achieved brilliantly by writers noted in the Bibliography such as Siegfried, de Madariaga, Blücher von Wahlstatt, or, most recently, by Barzini. Views are often strongly rooted: a French friend described the Belgian Walloons with great conviction as 'having all the

Frenchman's vices but none of his virtues'; most Germans view the neat and orderly Swiss as *spiessbürgerlich* (untranslatable, but a sort of smug, boring, rather narrow-minded 'square' mentality); most Slavs disdain the 'haughty Magyar', just as the French dislike 'la morgue anglaise'. Such perceptions unconsciously or otherwise colour international relations.

Over time the perception of national character may change. In the eighteenth century the Germans were seen as picturesque, preoccupied with philosophy and music, and apart from the Prussians, militarily inept. By the late nineteenth century they had come to be seen as efficient, expansive, and aggressive, while in the late twentieth century they are most commonly regarded as industrious and organizationally talented. At the same time, in the eighteenth and early nineteenth century, the English were considered turbulent and aggressive, but later in the century were viewed as industrially and commercially efficient and exceptionally hard-working, whereas by the late twentieth century a common external view sees them as indolent, unreliable, and a spent force living on past glories.

Nations owe in many instances a debt to an individual as a formative factor in nationhood. Apart from legendary figures in national iconography, modern nations have in many cases been the project of a great leader who fought and won for them recognition. In 1918 Mannerheim in Finland, Masaryk in Czechoslovakia, Pilsudski in Poland were key figures in the emergence of these nation-states; after 1945 Tito played the same role in Yugoslavia, skilfully piloting that ethnically complex country through a difficult phase in relations with the superpowers. In Africa we have the figures of Kenyatta in Kenya or Nkrumah in Ghana, and in India of Gandhi, but one of the most remarkable leaders at a time of national catastrophe was Atatürk in Turkey.

RACE

The term 'race' *must* refer only to physical anthropological characteristics: it is not to be confused with language, custom and usage, and the like, all of which are *ethnic* indicators. In this sense race has seldom been the primary element around which national identity has been moulded. Although in the blinkered society of the middle ages the Mongol troops struck fear in the heart of Europe by their sheer physical appearance, in most societies physical attributes have usually been less of a bar to integration than cultural or linguistic

20

traits. Sadly, where 'race' is a ground for discrimination and erection of social and economic barriers, the most common physical attribute invoked is skin colour.

In Europe the 'racial' types of Nordic, Dinaric-Alpine, and Mediterranean have been so intermixed from prehistoric times that physical attributes have played no role in the emergence of national identities. The farcical claims by the National Socialists in Germany of the superiority of the Nordics and their plans to breed a pure Nordic Germanic people represented an example, by no means unique, of the politicization of race. The rather synthetic nations of Latin America contain an amalgam of at least two racial elements: the European Iberian former colonial elite and the socially and economically disadvantaged Amerindians. The most successful amalgam has appeared in Brazil, where the Portuguese tradition has resulted in a multiracial society of European, Amerindian, and African elements with a distinct national identity, while a similar multiracial national identity has been forming in recent times under Castro in Cuba. In the United States, despite the remarkable success of integrating diverse European immigrants into a strong national identity, the other major racial elements of Afro-Americans (Blacks) and the Indians are still distanced by a serious social and economic divide. Real effort at integration and dismantling discrimination began effectively only after 1919, but that is perhaps not surprising when one considers the long relentless struggle with the Indians and the rigorous divide in national solidarity over the slave issue in the testing bitterness of the Civil War.

Just as pressure has been exerted against ethnic minorities, so some dominant elites have sought to prevent racial groups from sharing equally in the nation, with every obstacle put in the way of advancement, and even those who have managed to overcome these have been denied true equality with their peers. The most blatant contemporary example is South Africa, where Boer traditions among the politically dominant have applied strangely conservative and racialist views, nurtured by the Dutch Reformed Church. Races are kept apart irrespective of any cultural or linguistic considerations, defining where each may live and what civil rights they may exercise — the policy of Apartheid, which, though legally loosened, remains substantially in practice through economic and social pressures. For a long period, the 'White Australia' policy sought to make settlement by any but 'White Caucasians' exceedingly difficult, governed by a fear underlying most ethnic or racial relationships of being swamped by the other party. Australians feared that an open immigration policy would result in their being innundated by Indians and Chinese. Jealousy of the economic success of Indian immigrants in East Africa brought a policy of 'africanization' in Uganda, resulting in expulsion of Asians. In Iceland racial feelings went so far as to force the government even to debar Black servicemen on US military bases. In the Soviet Union integration of non-European groups in Central Asia has suffered less from racial bias than from the cultural force of the antipathy of Islam towards Slavs as Orthodox Christians.

SOME EXAMPLES OF HOW NATIONS EMERGE

The formation of every nation has arisen from a unique combination of factors, yet some themes are common, though occurring in many different guises. First, there is the recurrent theme of a fear of or a threat from outside pressures; second, there is the frequent case of a desire to be rid of foreign domination, the desire to be able to 'do one's own thing'; third, there is the play on the need for a separate identity because of 'being different'. Although there is usually a cultural identity and other factors in common among those drawn into a nation, providing a common iconography, this is not absolutely imperative, for the first two cases may surmount any differences in culture and other fields, if the threat is strong enough. There must, of course, be a leadership and a focal point to weld the nation together, with centripetal forces stronger than centrifugal pressures. The role of such leadership is to define the common bonds and provide a national iconography. Communion of interest and a common aim are, of course, vital and may surmount differences in the cultural sphere — this is important in making new nations from mixed origins: we find it well displayed in the national identity of the United States or Australia. It also typifies new nations stitched together from different indigenous groups in a territory, as in newly independent colonial territories of Africa, although in some of these countries there are such weak bonds that only the need to stay together because there is no acceptable alternative appears to give some semblance of nationhood.

The deepest foundation of the English nation lies in several small, unstable kingdoms sharing a common Anglo-Saxon tradition. Despite fluctuating fortunes, especially in the long struggles with the Norse, the Anglo-Saxon kings came to accept the overlordship of Wessex. Under Edward the Confessor nearly all the Saxon lands were united, but it was disagreement over his successor that gave the Normans the opportunity to invade and conquer, subsequently establishing a consolidated English monarchy. The Norman aristocracy introduced French feudalism and institutions, but the Saxons kept much of their language, some of their institutions and patterns of thought, while Norman and Saxon first came together in the church and slowly merged into a new English identity. Most important was the emergence by the fourteenth century of a robust English language, basically Anglo-Saxon with powerful French influence. The feeling of the English to be a nation was strengthened by the long wars over conflicting territorial claims between the French kings and the Angevin kings of England. In the sixteenth century the Reformation further cemented national identity, marking a clear rejection of foreign domination, epitomized in the crisis of the Spanish-Armada invasion threat. Tudor despotism strengthened the landowning and merchant classes, encouraging adventures overseas that fired the spirit and imagination of the nation. Loyalty shifted from the old vertical allegiance of feudal times to a lateral allegiance between the populace.

The Tudors, with a firm hold on England, centralized government, insisting on the primacy of English in their Welsh lands, though preserving Welsh identity through the crown prince as Prince of Wales. Scotland remained apart, always under pressure to pay homage to the English Crown. Scottish identity emerged through the struggle to keep English power at bay, though the Scots were often deeply divided and thus weakened in their ability to withstand constant English pressure. Scotland was given an effective ruler in James VI, whose prestige and power increased when he also became James I of England. The final Act of Union in 1707, though not popular, opened the way to economic advantages within the large English mercantilist trading sphere, yet at the same time preserved the Scottish law and legal system as well as the church. The United Kingdom comprised one state but brought three nations into a wider British nationhood without entirely depriving them of their separate identities.

The idea of a French nation grew as the kings of France, from their seat in Paris and the Ile de France, spread their rule, winning allegiance of other rulers, like the powerful Burgundians. In spreading their control southwards, they took lands then of different language and culture, justifying their action as a crusade against the heresy of Albi. The passions aroused in the long struggle against the domination by the English kings of much of western France helped cement French identity and build part of a national iconography through the story of Jeanne d'Arc and other events. By the fifteenth century the centralizing policies of the French kings (notably Louis XI), in step with skilful diplomacy, spread a common language in the dialect of northern France, which early in the sixteenth century was made the sole official language as a cold-blooded policy to ensure its dominance drew the nation ever closer together. The French kings allowed no regional variations as did the English kings in Wales and Scotland, forging a standardized and centralized nation of remarkable coherence and unity.

A vague sense of a common German identity was imparted from the fifteenth century in the title 'Holy Roman Empire of the German Nation', though the Empire was really no more than a collection of small, diverse sovereign territories with a tolerable measure of homogeneity in language and custom that allowed people to identify together as Germans, even though their first loyalty was to their *Stamm*. Sadly the brilliant medieval colonization eastwards and southeastwards, extending the area of German language and culture, did not contribute to a closer national and political unity. The Reformation did, however, provide a step towards nationhood through acceptance of the Lutheran translation of the Bible as the standard written language. The subsequent religious wars, so often with foreign intervention, further cemented the feeling of nation, though they did nothing to encourage political unity. The small, particularist states were happily German, but saw no need to surrender sovereignty to the wider unity of a political nation. A further bolster to national sentiment came with the Napoleonic period, encouraged by the ideals of the French Revolution but crystallized by resentment of French domination. Though the Congress of Vienna greatly simplified the excessively fragmented political map of Germany, no subsequent move towards a single political unity ensued: as Metternich had said in 1815, Germany remained 'a mere geographical expression'. The newly instituted Confederation was little less of a fiction than the defunct Empire: the real urge to unification came with the upsurge in economic life through the *Zollverein* of 1834. The revolutions of 1848 and the competition for leadership between Hohenzollern Prussia and Habsburg Austria achieved little, other than to exclude Austria from coming within any political unity. That unity was achieved finally in 1871, in the 'Little German' form, when Prussia took leadership of the Second *Reich*, in which it had two-thirds of the population and the dominant economic and military strength. The Second *Reich* had sufficient prestige, power, momentum, and symbolism to attract recognition of there being one German people, even if local loyalties remained strong. That single identity was further cemented in the brief centralized nation state of the Third *Reich* between 1933 and 1945. Perhaps we need to recognize the Germans' ability to live in separate political entities and yet retain the feelings and bonds of a single German people, *Deutschtum*, as part explanation at least of the apparently successful separate existence of two sovereign German states since the Second World War.

Italians were aware of a hazy national identity well before there was a single Italian nation state. The Congress of Vienna perpetuated the pattern of small states, most to some measure under Habsburg domination even though ruled by Italians, while the Papacy supported the absolutist rule resented by most Italians. Leadership for nationhood came from the urban professional classes and some aristocracy. Uprisings in the Revolution of 1848 were contained and the resented Austrian domination and absolutist rule returned, but in little over twenty years Piedmont (the leading state) diplomacy, besides more armed uprisings and the struggle for supremacy between France and Austria, brought Italian nationhood. As so often in national struggles, much was owed to one individual, Count Cavour, who saw the rise of Piedmont translated into a unified independent Italian kingdom, to which all Italians would be willing to be loyal. Occupation of Rome in 1870 marked completion of a united Italy as kingdom and nation, though the Piedmont prime minister thought that, having made the kingdom, the next task was to make Italians! Though plagued by powerful regional tensions and conflicts of interest between different groups, these centrifugal tendencies have not been sufficient to break the centripetalism of a nation believing in an iconography rooted in the tradition of Roman civilization.

The Magyars arrived in Europe as nomads in the ninth century but quickly became sedentary, nevertheless retaining their Uralic language, and believing themselves superior to the peoples around them. Before domination in the sixteenth century by the Turks, the legal status of belonging to the Magyar 'nation' was limited to 'nobility' (not all rich and some even without Magyar speech). After the Turks were driven out in the late seventeenth century, many areas were resettled by non-Magyars under Habsburg orders, but during the following century Magyar influence increased and Hungary was regarded as a separate kingdom (largely the prize for support of the Habsburg against Prussia), albeit under rigorous Viennese control. The eighteenth century witnessed rising pressure for Magyar to replace Latin or German in public affairs, particularly as the language was used by all classes and had its own literature, even though it was spoken as a mother tongue by less than half the population of the kingdom (Fig. 1.5). Membership of the nation now shifted from class to language and culture and an effort was made to magyarize other

25

Figure 1.5 Ethnic composition of the Hungarian Kingdom within the Habsburg Empire

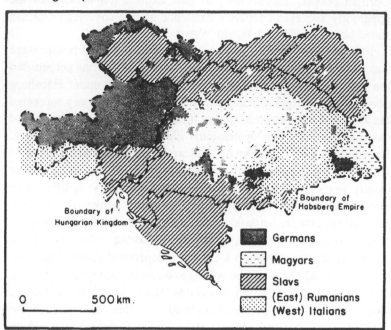

Adapted from *Grosser Historischer Weltatlas*, vol. III, Munich 1981

groups in the kingdom, who were regarded as 'nationalities' rather than 'nations', but their response was to take a greater interest in their own language and culture and to seek closer ties with their kith and kin across the kingdom's boundaries, alarming the Magyars, who suspected their loyalty. Influenced by the Revolutions of 1848, Hungarians demanded more freedom from Vienna and Magyar national awareness intensified, but not until 1867, after the Habsburg had been humiliated by the Prussians, was a limited sovereignty accorded in the 'Dual Monarchy'. Magyar nationalism insisted on ruthless magyarization within Hungarian borders, but it was unsuccessful and served only to isolate the 'Haughty Magyar', making only enemies of others, so after 1919 non-Magyar peoples showed no compassion as Hungary's borders were drawn so tightly that 'the pips squeaked'.

Rumanian national identity was fired by a powerful but rather liberally interpreted iconography, claiming direct descent in language and tradition from the ephemeral Roman presence in Dacia, as well as stimulating a keen attachment to territory, the *Ţara Românească*. Several interpretations of how the Romance element (now seen only in language) survived a turbulent stream of Slav and other invaders have been made. The early nineteenth-century removal by the Turks of the Phanariot Greeks who ruled Moldavia and Wallachia for them, but had become too powerful, gave Rumanians a chance to fill vacant places in administration, education, and trade. Further national development was retarded by great power machinations, though a true Rumanian nation-state appeared in 1859, the 'Old Kingdom', tributary to the *Porte* until 1878 and formerly recognized in 1881.

Driven from their homeland in Roman times, the Jews were scattered by the diaspora and yet, through their cultural and religious distinctiveness, preserved their identity as a people, helped by living in close communities and involvement in trades like money-lending (repugnant to Christians), even though put in a position vulnerable to dislike and mistrust, often spilling over into open persecution. Jewish communities became markedly stamped by influences of the people among whom they lived, so those of East Central Europe, attracted there by the liberal policy of the Polish kings, diverged in many respects (notably in the wide use of Yiddish) from Mediterranean Jewish communities with powerful Iberian cultural influences. As political, social, and economic constraints under which the predominantly urban Jewish communities lived eased in the nineteenth century, Jewish society diversified, often drifting away from the older orthodoxy, and many members began to regard themselves as members of the nations among whom they lived.

Rising pressures among peoples in East Central Europe for national recognition late last century encouraged Jewish endeavour to win back their homeland, a feeling given added force by resurgent antisemitism. Migration of Jews to Palestine to re-establish a national home was encouraged by Herzl's *Zionism*, while the polyglot migration generated a desire for a common language. Yiddish was not considered suitable, so Hebrew was developed into a modern spoken form. Zionism initially received little enthusiasm from Jews in Western Europe and America, many of whom felt their first loyalty to their adopted nation. Rising Jewish numbers in Palestine quickly brought conflict with a well-established Arab population, but Jewish migration continued to swell in the Great Depression of 1929–31 and as Hitler encouraged antisemitism. After the Second World War the United Nations sought to establish a Jewish national state, which appeared when the British withdrew from Palestine in 1948 and quickly consolidated itself. The state of Israel has witnessed emergence of one of the most assertive

27

and self-conscious national identities in the world, but has successfully withstood the wrath of the Arab world.

Australia has developed its national identity among descendants of an immigrant population, largely from the British Isles, with some more varied recent immigrants. The new nation has emerged in a much freer social milieu with a greatly enhanced social mobility, set in an entirely different environmental framework, so Australians have grown apart from Britain (even though many institutions and customs have been retained or adapted) and are now recognizably different. The Statute of Westminster of 1931 gave the 'White Dominions' the right to manage their own affairs and to assert their own identity, while the Second World War showed Australia had to be able to look after itself and chart its own course as the imperial ties with Britain were dismantled, all providing an impetus to sculpt its own identity.

The American nation arose from an amalgam of polyglot and culturally diverse but mostly European immigrants, seeking in common the opportunities of a new environment and an escape from the fetters of past experience. The germ of the nation lay in the original, principally British, settlers who, through the insensitivity of the home government and their tendency to grow apart from their compatriots, came into conflict with Britain during which they won independence. Having defined the constitutional values, aims, and institutions for the nation, they did not consciously form an elite and admitted European immigrants as equals. Assimilation and integration into American society were quick, certainly within two generations, for the circumstances driving most migrants from their homeland gave little desire to retain ties with it. The economic success of America, giving rising political power, was a matter of pride that encouraged loyalty and identity. Nevertheless for many migrants, especially those in the ethnic 'ghettoes' of cities like New York, there was sometimes a struggle between the Americanizing influences of school and workplace and their own identity mirrored by their church, cultural associations, or local newspaper in their own language.

The steamship and the railway made possible the rising flood of immigrants, so that by late last century pressures to restrict the inflow began, not only from 'established' Americans but even from quite recent incomers, the characteristic reaction of an established population to a fear of being swamped by 'strangers'. Immigration peaked in the early years of this century and quotas for numbers from defined areas were imposed after the First World War, weighted against Eastern and Southern Europeans, the last major flood, and especially against non-Europeans. Neither the native Indians nor the

Afro-American ex-slave population were given much chance other than to accept American identity on a somewhat second class basis. Later generations of Euro-Americans have tended to a sentimental attachment to the origins of their forbears, particularly marked among some groups like the Irish and the Scots, while recently Afro-Americans have also been stirred to an interest in their African inheritance. The effect has on occasions been for pressure groups to use their ethnic origins (as among Poles or Italians) to seek to influence United States foreign policy. Such intimate links could have provided United States foreign policy with diplomatic and political skills of an exceptional order in international relations, but unfortunately there is little evidence of that.

Sometimes a territory gains independent political status without a genuine national sentiment existing within it. We have already seen the difficulty of generating a true nation between Walloon and Fleming in Belgium, a state arbitrarily created in the wake of the Napoleonic Wars. A small vociferous group, often drawn from only one part of a diverse population, has in many instances managed to win political independence from an imperial system and has then faced the task of drawing together disparate elements into nationhood. Nigeria illustrates this problem, a large territory glued together by the British in the nineteenth century, but filled by peoples greatly diverse in culture and language, with only fifty of its four hundred languages of standing enough to be used in radio broadcasts. As a federation after independence, it has not managed to draw its peoples together in real national coherence, with the bitter Ibo secession and genocide-like Biafran War in 1967. The first threefold federal division was reorganized into a twelvefold division, but even so has not overcome a fear by many groups of domination by powerful peoples like the Yoruba or Ibo or the northern Fulani and Hausa.

The concept of the nation as the primary political force in its own independent territorial state received a major boost after the First World War in the Wilsonian doctrine of self-determination. The idea was that each nation (rather open-ended in definition) should decide its own future and manage its own affairs in its own *de jure* territory. The doctrine was applied rather unequally as, for example, in Czechoslovakia. Czechs and Slovaks uneasily united but, because their nation-state was defined within historical boundaries, they found large German, Magyar, and Ukrainian minorities within it. These minorities were never consulted about whether they wished to be within the Czechoslovak nation-state — a clear denial of the 'self-determination' accorded the Czechoslovaks. National expression of self-determination as the basis for political recognition and the institution of a territorial state has been accepted with reluctance by many governments faced by vociferous minorities. The decade 1940–50 saw brutal means used to simplify the ethnic quilt of Europe to provide uninational state territories, but here as elsewhere it was not completely achieved. Since the 1960s many small nationalities have vociferously demanded greater autonomy if not independence. Certainly evidence suggests both multinational states and very small nation-states can be successful, but also equally unsuccessful. In the multinational state it is the attitude of the several groups towards each other and the desire to remain together that are crucial, though economic viability is important. The latter is perhaps the most crucial of all factors in the success of the small uninational state, which inevitably depends on the economic goodwill and tolerance of otherwise powerful neighbours. The excessive territorial fragmentation that made Germany before 1815 so politically impotent and economically weak is surely a warning. In Britain, if greater autonomy were conceded to the Scots and Welsh as 'nations' within the British framework, it is not unlikely eventually similar claims would come from Northumbria (the scenario is easy to imagine) or from Cornwall, for which not unreasonable cases could be made. In 1945 Bavarians sought political independence from the wreckage of the *Reich*, an aspiration based on the model of Austrian nationhood, although on the other hand *Deutschösterreich* in 1918 had sought attachment to Germany, for which a cogent argument was also made.

FORMAL NATIONALITY

Political geographers must be aware of the distinction between 'ethnic' and 'legal' nationality. Every one of us is born without choice into a group whose *ethnic* nationality we inherit, but we also have a *legal* nationality which may or may not be the same, formally defined by law. Legal nationality may depend on place of birth, on our parents' legal nationality, or even by our applying to be a citizen of a particular state, known as 'naturalization'. If our legal nationality is withdrawn or we reject it and can find no other state to accept us, we become 'stateless', a most unfortunate condition. Entitlement conditions to a particular legal nationality can be complex. *Ethnic* nationality is a matter to a large degree of self-assessment, what we feel ourselves to be. Certainly factors like 'mother tongue' and family tradition will be most influential, but some people find themselves

30

drawn between two or more groups (e.g. children of mixed marriages living in a third country). For official purposes of measurement it is extremely hard to find reasonably infallible criteria and even harder to get completely honest answers, particularly from minority groups.

2

The state

The state is a sovereign legal entity, internationally recognized, whose responsibility is to organize and guarantee the welfare and security of its citizens within its territory, where it is the supreme authority, tolerating no competitor or challenge to its sovereignty and demanding obedience from the inhabitants. [The form 'state' refers to the nation-state; members of a federation so designated are given the title in italic.] The state holds the sole right to make general arrangements for the conditions of life of its citizens, as well as to enforce their observance. Acceptance of this, whether by persuasion or coercion, by the overwhelming majority of its citizens gives the state legitimacy and effectiveness. Constitutional legal definition of the state's competency, whether in written form or by precedent, lays down its nature and from whom power emanates. It is also customary to describe the structure of the state apparatus and the role, responsibility, and power of such key elements as the legislature, executive, and judiciary. This is the basis for recognition of the state by the corresponding competencies in other states of the international community. The state is, of course, not unchanging and there is generally a well-defined legal procedure for amendment, usually designed to prevent irresponsible or precipitate action.

Management of the state apparatus is the task of government which plots its course through policy decisions. Change of government, usually bringing change of policy, does not affect the sovereignty or legitimacy of the state. The nature and structure of the state may, however, undergo radical change, as in the complete revision following the Bolshevik Revolution in 1917 in Russia or the creation of a federal republic in the German *Grundgesetz* of 1949. New governments, especially where they come to power through a *coup d'état* or revolution, may not be recognized by other governments, though this does not normally influence the sovereignty and legitimacy of the state as such. Nevertheless non-recognition may have significant if temporary geographical effect, disrupting channels of trade and other international links, especially if formal sanctions are invoked. The Unilateral Declaration of Independence by the Smith government in Rhodesia did not affect the sovereignty or legitimacy of the Rhodesian state in legal or geographical terms, though African trade and other links were recast by sanctions against the government itself. The aim of sanctions is to make a government's position untenable, so forcing it to change course or dissolve. The state apparatus is more enduring than governments themselves, often affecting or constraining the policies they may pursue; for example, it is commonly said the civil service and not the government is the real power in France.

Apart from unfortunate people made stateless by political vicissitudes, every person is born a citizen (or elects to become one by 'naturalization' (p. 30)) of a sovereign state or its dependencies. Citizenship involves both rights and duties in return for the state's protection, while the more liberal constitutions allow citizens, in return for some surrender of freedom and carrying out certain obligations, a measure of influence over government policy through democratic participation. The control exercised over individuals may have quite notable geographical consequences; for example, constraints on the freedom of internal movement in the Soviet Union affect patterns of settlement and transport, while international movement of people is much influenced by the travel and residential legislation of states governing their own citizens or visitors.

To exercise sovereignty, every state must have control over both people and territory. In the organization of the state apparatus, there is a powerful territorial element: nineteenth-century France was highly centralized on Paris, while Switzerland was a liberal federal structure and consequently greatly decentralized, and the impact of this structure can be clearly traced in the geographical pattern of such elements as the transport system or the urban hierarchy. Most recently some states have sought to give the apparatus at least an appearance of decentralization. During the economic difficulties of the 1960s in France, effort was made to decentralize by spreading functions to provincial centres, countering the claim of 'Paris et le désert français', whereas in the German Federal Republic the state apparatus has been extensively decentralized from its inception in 1949.

FEDERATION

The unitary state is the most common organizational structure, comprising some 95 per cent of the total, though controlling only half the land area. The remaining 5 per cent (some fifteen States) form a special case in several different formats of particular interest to geographers, the federal structure: classic examples are the United States, Soviet Union, Switzerland, Canada, and Australia (Fig. 2.1). Federalism seeks to surmount one of the most difficult tasks of any state, the maintenance of cohesion by preserving the balance between the centripetal forces drawing it together and the centrifugal ones pulling it apart. Federalism functions best where the desire for it is mutual among the different regional communities, seeking a union for their common advantage but without losing their own identities, as could happen in a unitary state. The federal state (in recognizing regional diversity) is appropriate where regional heterogeneity is pronounced and can be territorially defined within the national territory. It replaces a single uniform national loyalty by equalized regional loyalties acting in concert for national solidarity. Although regional diversity does exist in many unitary states, it is commonly not pronounced enough to stimulate sufficient feeling among the population for a federal structure. Territorial factors appearing to stimulate pressure for federalism include ethnic or cultural diversity, strong and usually long-standing particularist loyalties, but also the dimensional element affecting centralized management of exceedingly large states or those where physical conditions (especially in the past) have made interregional communication awkward.

Whereas the unitary state administers all its functions and responsibilities centrally, leaving only the basic everyday tasks to local government, the federal state lets only the highest echelon of responsibilities rest with central authority, such as foreign policy and defence. The *states* in a federation manage the bulk of the middle and upper tier of responsibility themselves, thus becomimg the equal of the central authority, influencing formulation of national grand strategy, with their autonomy and functions defined and guaranteed constitutionally.

A federation is often claimed to be a flexible territorial organization, able to create new *states* by fusing together or dividing already extant territories, perhaps creating entirely new units. Switzerland provides examples of the division of the Canton Basel into two in 1833 and the recent secession of Canton Jura from Canton Bern in 1979. Elevation of territories to the rank of federal *states* may be cited

Figure 2.1 Growth of the Swiss Federation — dates of foundation of the Cantons.

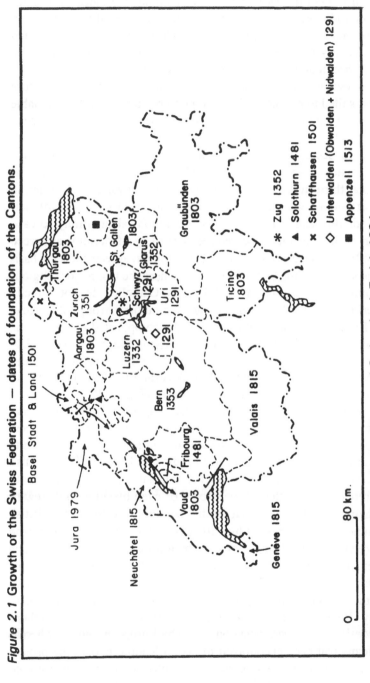

Basel Stadt & Land 1501

Thurgau 1803

Zurich 1351

St. Gallen 1803

Schwyz 1291

Glarus 1352

Uri 1291

Aargau 1803

Graubünden 1803

Luzern 1332

Ticino 1803

Bern 1353

Valais 1815

Fribourg 1481

Jura 1979

Neuchâtel 1815

Vaud 1803

Genève 1815

✳ Zug 1352
▲ Solothurn 1481
✕ Schaffhausen 1501
◇ Unterwalden (Obwalden + Nidwalden) 1291
■ Appenzell 1513

0   80 km.

Compiled from Fahrni, D., *An Outline History of Switzerland*, Zurich 1984 in the addition of Hawaii and Alaska as *states* of the Union in the USA in 1959 or the entry of Newfoundland into the Canadian Federation in 1949 (Fig. 2.2). A federal structure seemed to be the only hope of holding Nigeria together on its independence in 1960, although subsequently this extremely complex territory was recast from the initial threefold division into twelve federal *states*.

The evidence is as inconclusive as to whether a federal structure is best suited to new countries seeking to build an identity acceptable to diverse regional interests, as it is on the question of its being most appropriate for very large territories. The origins of federal nation-states do seem, however, to follow a number of distinct paths. Federalism has worked best where adopted by mutual agreement: the original thirteen American colonies saw the advantage of acting in concert and yet were anxious to maintain as much as possible of their own identity and, with their close geographical proximity to each other, their objectives appeared best served by a federation as agreed at the Philadelphia Convention of 1787. Mutual agreement is the basis of the success of the Swiss federation, where the two major religious denominations, Protestant and Roman Catholic, do not directly superimpose over the two main linguistic-cultural groups of German and French in a double alliance, but the complicated pattern of the resultant bonds is seen as a significant unifying factor. Mutual realization of their common destiny drew the highly particularist *states* together into the Commonwealth of Australia in 1901, though with some reluctance and no great enthusiasm (Fig. 2.3).

Federalism is often a compromise, especially where different peoples with all the qualities of nations lack the strength, resources, or even will to become independent nation-states and instead unite in some federal arrangement, as in India and Nigeria. There is also the case where federalism has been an imposition, established without reference to the wishes of the people or even blatantly in disregard of them, as in the two unsuccessful examples created in the British Commonwealth, whose failure arose from the lack of a sound *raison d'être* and through being composed by units out of balance with each other. The Central African Federation of 1953 comprised the self-governing colony of Southern Rhodesia and the protectorates of Northern Rhodesia and Nyasaland, where only economic complementarity linked three otherwise incompatible territories. Southern Rhodesia had a government dominated by European settlers, whereas Africans had been encouraged to take a leading part in the protectorate governments, posing serious administrative problems for concerted federalism. At the same time, in the electorate, Southern

Figure 2.2 Territorial growth of the USA.

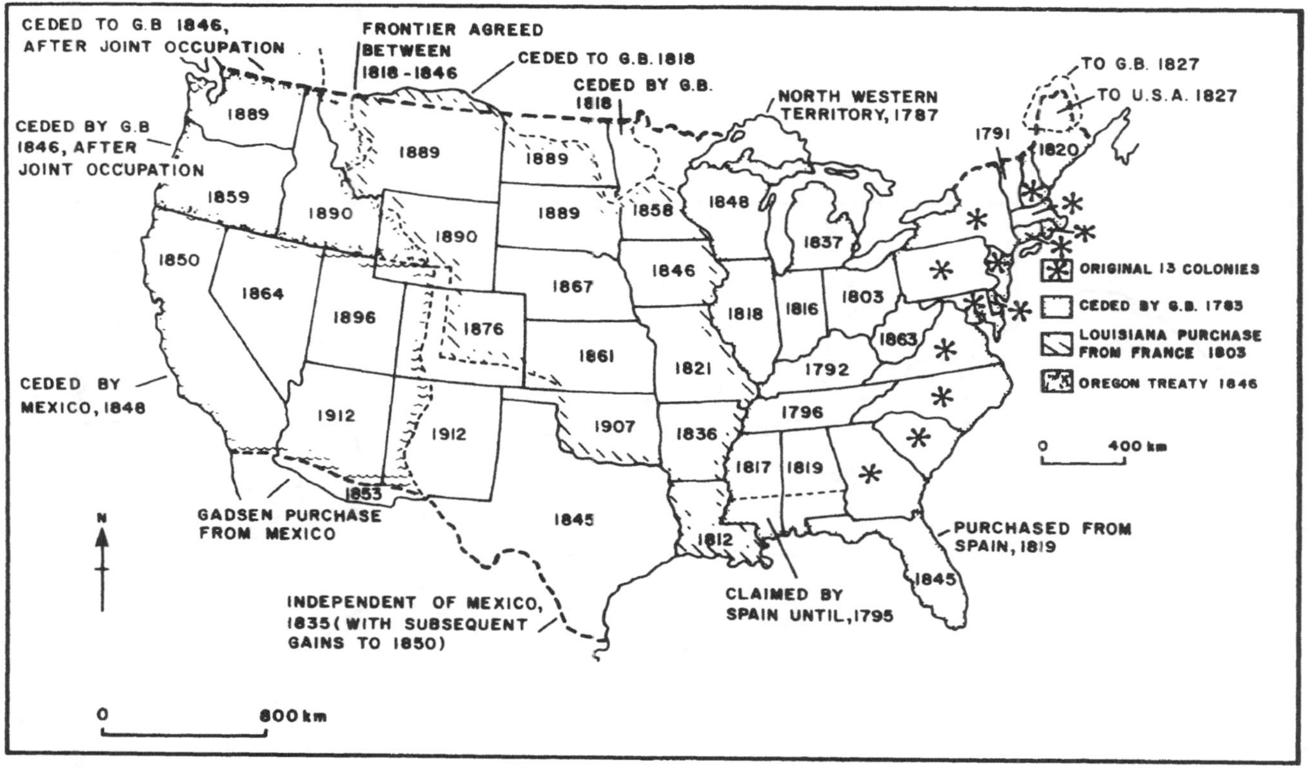

CEDED TO G.B 1846,
AFTER JOINT OCCUPATION

CEDED BY G.B
1846, AFTER
JOINT OCCUPATION

CEDED BY
MEXICO, 1848

FRONTIER AGREED
BETWEEN
1818-1846

CEDED TO G.B.1818

CEDED BY G.B.
1818

NORTH WESTERN
TERRITORY, 1787

TO G.B. 1827

TO U.S.A. 1827

1820

1791

ORIGINAL 13 COLONIES

CEDED BY G.B. 1783

LOUISIANA PURCHASE
FROM FRANCE 1803

OREGON TREATY 1846

400 km

0

PURCHASED FROM
SPAIN, 1819

CLAIMED BY
SPAIN UNTIL, 1795

INDEPENDENT OF MEXICO,
1835 (WITH SUBSEQUENT
GAINS TO 1850)

GADSEN PURCHASE
FROM MEXICO

1889

1889

1890

1890

1876

1912

1853

1896

1912

1864

1859

1850

N

800 km

0

1803

1863

1837

1816

1792

1796

1818

1821

1836

1819

1817

1812

1845

1848

1858

1846

1861

1907

1867

1889

1889

1845

Compiled from various sources

Figure 2.3 Emergence of the Australian Commonwealth.

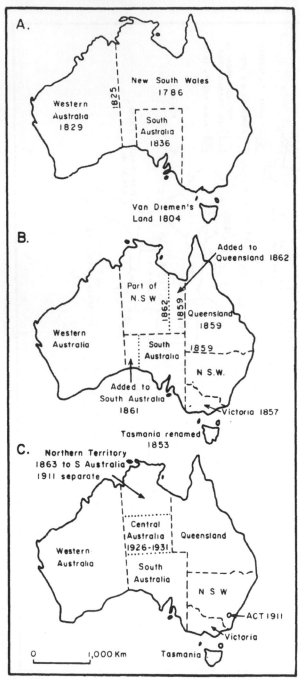

Adapted from de Blij, H., *Systematic Political Geography*, New York 1967

Rhodesia predominated, regarding the other two units as dependencies. European interests held three-quarters of the seats in the federal assembly, so the African majority population felt little confidence in the federation safeguarding their interests. The federation fell apart when the protectorates seceded in 1962–3. Even less good reason underlay the Federation of the West Indies, conceived principally in the image of its major units, Jamaica and Trinidad, which dominated the many small island members. Long term, some federal structure may re-emerge through economic necessity with common problems of urban development and employment among the weak island economies.

Differences between constitutional statements and administrative reality may produce a pseudo-federalism in an otherwise centralized state. The Soviet Union is constitutionally organized as a union of socialist republics and appears on paper as an admirable federal structure. In reality the one-party political system, the centrally planned economy, and massive bureaucracy seated in Moscow preclude effective federalism. Nevertheless some modest autonomy is allowed republics and some lower orders of 'autonomous' units, but even so the federation is dominated by the massive Russian Republic, the RSFSR, in population, wealth, and resources. Also in this category, Yugoslavia emerged in 1918, drawing together Slav peoples of markedly different background and experience, unified only in their desire to be free of foreign overlords. The essential weakness has consistently lain in the suspicion and competition between the more wealthy and economically developed Croats in the north and the Serbs in the south, who gained their independence earlier and thus believe themselves to be the rightful leaders of the community of South Slav peoples. This attitude was strengthened by having provided the king in the newly created kingdom of 1918, but such a unitary nation-state did not satisfy the aspirations of other peoples and by 1939 some tentative moves towards federalism had been initiated. In 1945 a highly centralized, bureaucratic, one-party state on the Soviet model was established, effectively federal only in name, but subsequently some effort was made to bring the territorial-administrative responsibilities more in line with cultural and linguistic divisions, but meagre concessions to regional autonomy did not satisfy aspirations. Demands for further concessions during the 1960s and 1970s made little progress as central government and the Party lost interest in furthering federalism and turned back to a more rigorous centralist policy.

Even apparently unitary states may exhibit some federalist traits. In the 'United Kingdom' of Great Britain and Northern Ireland,

Scotland has retained its own educational and legal system, among other small concessions, and some measure of self-government existed in Northern Ireland until the early 1970s. Wales has been represented by the heir to the throne as Prince of Wales and more recently the special nature of the Principality has been acknowledged by establishment of the Welsh Office. Rather special forms of local autonomy, historically conditioned, exist in the Isle of Man and in the Channel Isles.

CONFEDERATION

The terms *federation* and *confederation* ought not to be interchanged. The difference is well expressed in German between *'Bundesstaat'* and *'Staatenbund'*. A *federation* is created by internal constitutional means in dividing a sovereign state into federal components, guaranteeing their autonomy and defining their responsibilities and those of central government, thus sharing the authority of the state between two orders of competency. A *confederation* is established by international agreement between sovereign states, in which some carefully defined powers are surrendered to a central body for specific goals. The central body comprises representatives sent by the participating states but not elected by popular vote.

Confederations have not generally had a successful history, tending either to merge into closer associations or to disintegrate. Napoleon created the ephemeral Confederation of the Rhine in 1806 and it was swept away in 1813, while the German Confederation created by the Congress of Vienna in 1815 had an unhappy, impotent history until its collapse in 1866. The Confederal Diet in permanent session at Frankfurt am Main, under the permanent presidency of Austria, comprised delegates from all German sovereign states, but it was seldom able to get the unanimous decisions needed to enforce its deliberations. In its fifty-year history it never even managed to achieve its primary purpose of establishing a confederate army. The end came with the war of 1866, precipitated by the struggle between Prussia and Austria for leadership of Germany. Defeat of Austria left Prussia supreme, so the old Confederation was abolished, replaced by a North German Confederation completely under Prussian domination, which left the South German States to decide their own relations to it, while Austria was excluded. This was only a step towards the creation of the Second *Reich* (1871), an empire of a vaguely confederal nature firmly under Prussian management.

The early history of the United States of America illustrates well the difference between the federal and confederal principle. During the War of Independence each colony had amended its charter into a constitution to serve a sovereign state, but there was a need for them to have a closer and more permanent union. The task was undertaken at the second Continental Congress, which drew up the so-called Articles of Confederation in 1777 (not finally ratified by all colonies until 1781) as a voluntary framework, resting on delegation of sovereign powers of a general nature to a central authority, but withholding the two vital powers of the right to raise taxes and mint money and to regulate commerce, and there was no provision for a common foreign policy. It did nevertheless solve the issue of land west of the Appalachians, allowing setting up new states which could eventually join the Confederation. Once the threat of war receded, the weakness of the Confederation was revealed as members returned to their own individualistic, penny-pinching ways, acting in self-interest to the detriment of the others, for the co-ordination of action was too vaguely defined.

The problems facing the members brought a Federal Convention at Philadelphia in 1787 which drew up an entirely new form of government, seeking to reconcile local control exercised by the thirteen semi-independent territories and the power of central government. The powers of the national government were to be carefully and specifically defined: all other powers and functions were to be understood as belonging to the *states*. It was appreciated that central government needed real power, previously lacking, so it was empowered — among other things — to control money supply by minting coinage, to regulate commerce by tariffs, to conduct foreign policy, and to arrange defence. The search for an acceptable balance and interrelationship between local and central power was achieved in central government by a division of power in which one house was for representation equally among the *states*, the other for proportional representation of the *states* according to their population. The written Constitution of 1788 wisely allowed for amendment, particularly important as the area of the United States expanded and developed.

THE GEOGRAPHICAL DEVELOPMENT OF THE STATE — SOME EXAMPLES

Changes in the nature and organization of the state within its territory over time have consequential implications for geographers. The state apparatus seems to have emerged as sedentary farming produced a surplus for exchange, allowing for trade, occupational specialization, and an increasing stratification of society. Evidence from the Middle East, the Indus basin, and the Nile valley suggests the growth of trade and specialization stimulated agglomeration in specialized communities, the early towns. With formal ownership or holding of land outside the ruling elite and with trading through markets, there arose a need for skilled persons to formulate and apply laws as well as to collect tribute or taxes. Formally appointed administrators managed and maximized the advantages of the social and economic system on behalf of the commonly hereditary ruling class. Such administration was especially necessary in societies where farming depended on exact irrigation management, as in all three early cradles, though the more advanced system of Mesopotamia probably arose from the more complex water management there. Religion appears to have played a significant role in development of the state apparatus, with major seats of administration usually in priestly centres, perhaps because rulers so often claimed descent from or empowerment by deities in order to legitimate their sovereignty. The priesthood, by its nature and experience, could supply capable administrators, able to impose their will by demanding obedience to avoid suffering the displeasure of the gods. Sovereignty was expressed as the exclusive right to management of a defined territory, whose inhabitants were absolute subjects of the ruler, who in return for their obedience organized defence through some form of military system and ensured the favour of the gods.

The city-state appears to be the earliest form, but population pressure or even greed stimulated some to expand their influence by war, political intrigue, or religious pressure, absorbing other territories. In Mesopotamia the rule by a single kinglike individual was practised from early in the third millennium BC and some powerful personalities united several city-states under their overlordship. As territories grew in extent, problems of movement and communication forced delegation of territorial authority to subordinates. In the early Mesopotamian civilization, four developments marked growing sophistication of the state apparatus. First, political theory and loyalty to the state developed. Second, as states grew in complexity, a strong bureaucracy and a professional army appeared, usually developed out of the royal household and stationed at key places throughout the domain. Third, the techniques of administration improved, with the development of writing as a key to communication and record keeping, easing management of larger, more complex territories. Fourth, the growth of trade brought a merchant class, providing an intermediary between the ruling elite and the peasant masses. As merchants were to a considerable degree dependent on 'government contracts', they were notably loyal to it, while they were favoured by it, because trade was essential to supply items not available at home (particularly for the military and for the royal household and elite) and offered wealth for taxation.

The Greeks organized their life around city-states (perhaps a pattern favoured by the mountainous terrain) which often united separated plains (as in Athens and Sparta), although in other instances fertile plains were divided among several political units (e.g. Boeotia and Attica), with social rather than geographical conditions generally the determinant. As in early Mesopotamia, application of Christaller's technique to define an early state module does not work very well other than in most general terms (Renfrew 1975, 1977). A rich, leisured class of aristocrats and administrators ruling a broad but relatively poor class of artisans and peasants marked Mesopotamia, but in Greece the aristocracy lost its power and kingships ceased to be hereditary. McNeil (1964) considers this an outcome of changing military tactics, with a shift from cavalry (whose cost favoured the aristocracy) to heavily armed infantry, the hoplites, in the seventh and sixth centuries BC. The hoplite phalanx through its military power and social cohesion brought change as ordinary townsmen or farmers became the key to military success, so that differences in wealth or social status lost their meaning, generating a spirit of egalitarianism and civic solidarity. The larger the phalanx the stronger the defence of the state, consequently everything was done to enhance the number and status of citizens; the Spartan response was to produce a classless, military society and adopt a marked xenophobia, whereas the Athenian response was a revision of its law and economic system to strengthen citizenship and membership of the phalanx. The power of the hoplites forced the aristocracy to surrender a major say in running the affairs of the *polis* to them, the foundation of democracy. In later times some of the more powerful city-states built empires by absorbing weaker members through territorial conquest, so that the 'classical' democracy of the city-state, where all citizens could attend the assembly, became hard to maintain, as in the Athenian empire where citizenship encompassed about 150 other cities. Though local democracy survived, the states themselves became ever more centralized and, after the time of Alexander the Great, rule by a single head, with a royal line of accession and often divine attributes, emerged.

Rome as a powerful city-state first conquered Italy and then spread into vast provinces around the Mediterranean and beyond. The Roman Republic was based on the Senate, dominated by an aristocratic class, though its base was ultimately extended to include consuls elected by assemblies of free citizens. The key to the Roman state was citizenship, defined by law rather than by territoriality: *civis romanus sum* — a Roman citizen enjoyed the status everywhere. The Roman state depended heavily on the landowners, whose estates were worked by indebted peasantry, a stratum of propertyless but free farmers. A similar stratum of *proletarii* inhabited the towns, while the economy in general depended on a large and growing stratum of slaves. Everything was regulated by an elaborate legal code, distinguishing between 'public' and 'private' responsibilities, rights and duties.

The spread of Roman territorial power made holding together a unified state increasingly difficult: the political geography of the Roman Empire was basically the problem of maintaining tight bonds but nevertheless balance between the core and the periphery. As the City of Rome itself grew in wealth and population, the dependence on the provinces for food and other supplies exacerbated the issue. Holding the far-flung provinces was made difficult through the most persistent of all imperial problems, finding adequate manpower for the military and for the administration. The Romans by necessity became masters of the psychological dimension of power, with the skilful use of the ominous presence of the legions in key positions, forcing loyalty on the hegemonic structure of the Imperium.

THE FEUDAL STATE

The feudal state emerged during the Dark Ages in Europe, though feudalism itself had several interpretations, and analogous systems also developed at differing periods in Asia. The 'classical' feudalism was established within the Holy Roman Empire, and combined characteristics of the Roman tradition and the looser usage of the Germanic peoples. Feudalism was in effect a system of protection in return for service or tribute. The upper levels of society gave military service to their lord or king; the lower strata paid mostly in labour and sometimes in military service. The ruler gave his most trusted subordinates grants of land and special privileges in return for loyalty and homage as vassals, an inheritance from the Germanic tradition. In the contemporary conditions of movement and communication, feudalism favoured territorial fragmentation, with power local and personal, but its often overlapping claims and powers aggravated conflict and friction. Feudal monarchs were *primus inter pares*, different in degree rather than kind from their lords, tied to them by reciprocal obligations. As the king was dependent on his most powerful lords for raising an army and gathering taxation, it was difficult for him to be truly sovereign rather than an unusually powerful suzerain. The lords commonly sought to hinder the king concentrating increasing power on his own person, so he and his retainers spent much time touring his domains in a show of strength, demanding frequent restatement from the lords of their homage. Some kings, as in England, were better able to control their barons, largely by carefully distributing their own estates and castles to retain a strategic advantage, a policy the Holy Roman Emperors failed to achieve. It was especially important to retain the absolute loyalty of the 'marcher lords' whose role was to guard the borders of the kingdom from hostile incursions, and who in consequence enjoyed special rights and privileges. Secular power was also frequently challenged by ambitious ecclesiastical rivals, powerful bishops holding large territories.

The feudal state was increasingly weakened by the rise of commerce focused on the towns, seen by Pirenne (1959) as 'islands in the sea of feudalism'. The growth of a monetary economy centred on the towns stimulated trade that undermined the feudal system based pre-eminently on subsistence. With charters and the right of self-government, besides a different social and economic structure, towns enjoyed rising power to challenge the monarch, who was often heavily in debt to urban financiers (e.g. the Fugger family of Augsburg). It was not surprising that towns in the Holy Roman Empire formed powerful political leagues and by 1489 the Imperial cities had become the Third Estate in the *Reichstag*. Kings soon discovered money could buy loyalty and military support, so that now land and people were taxable assets, what price service and vassalage? Though the declining feudal system shifted from service to monetary payments, peasant labour, useful if inefficient as a means of wealth creation, was often retained as a feudal obligation that died last in the lowest ranks of society.

The need of towns to control their own destiny encouraged them increasingly to assert their muscle in the state. Their experience of management through councils representing the various 'estates', such as merchants, clergy, and guildsmen, generated pressure for similar assemblies at higher levels. At the same time rulers were constantly challenged by the immense power of the wealthy church institutions, while the conflict between the Pope's claim to sovereign spiritual power and the secular rulers' claim for a similar temporal independence laid the foundation of the legal concept of sovereignty in the modern state. The strengthening of the individual temporal rulers at the expense of the universalist church also brought a closer identification of the people to their group, both in loyalty to their ruler and in attachment to their compatriots and their homeland.

THE ABSOLUTE STATE

As feudalism faltered, rulers strengthened their position, seeking more clearly defined, unified territories, while the old shifting loyalties were frowned on and law and order tightened. Better able to arrange their security, a continuity of policy and a calculable government for the rulers became possible. By the sixteenth century the view that kings had a 'divine right' to rule was propounded and the sovereignty of the ruler became absolute, expressed by the French king Louis XIV (1643–1715) as 'l'état c'est moi'. Kings ruled from their courts and not through them, while asemblies, such as they were, were seldom called and even less heeded. The state under the absolute Crown assumed a *dirigisme* in the economy, the mercantilist system, designed to maximize its own benefits. Relations with other rulers became an increasingly formalized system of diplomacy and dynasties sought to strengthen their influence by astute marriages.

The concentration of absolute sovereignty in the monarch and the consequent rigid *dirigisme* brought a reaction from commercial interests. Both the landed interests of the gentry and the commercial interests of urban craftsmen and merchants opposed mercantilism for the constraints it imposed on trade. As the commercial and agrarian interests came to dominate social and economic life, they exerted pressure for a greater involvement in the management of the state and the formulation of policy, to apply their own experience and ideas, introducing business methods, especially the concept of contract. The absolute state thus became the contractual state, where power was shared in a defined and constitutional form, recognizing at least the rights of the upper and middle ranks of society. Evolution of the constitutional state lasted from the seventeenth century in England to the late eighteenth century in France and even into the next century in Central Europe. In the late nineteenth and early twentieth century, the process was carried through to encompass all ranks of society and government had become fully secular, with church and state increasingly formally separated, while the sovereignty of the state was more sharply defined.

46

THE LIBERAL STATE

Out of the absolutist state developed the liberal state, less rigid and with a major function to safeguard the rights and liberties of its citizens. It also recognized free trade and the practices of commerce, whose concepts influenced its own organization. The state now became legally bound to recognize defined liberties, which could no longer be tampered with at the whim of the Crown or state. In order to fulfil its responsibilities to safeguard these liberties and other legal agreements, the state had to be strong, especially able to defend its citizens against outside threats, but it was now expected to interfere as little as possible in the daily life of the nation, especially to keep out of economic affairs, leaving market forces with the greatest possible play.

TOWARDS THE WELFARE STATE

Industrialization and the rising imperial frictions in nineteenth-century Europe weakened the concept of the liberal, *laissez-faire* state. Feeling arose for a greater involvement in planning and policy-making on behalf of the nation as a whole, a widely supported trend towards an interventionist and collectivist state. Governmental and entrepreneurial groups believed it would enhance national efficiency, while the broad mass of the population saw in it a possible way of improving living conditions. Under this influence the 'welfare state' began to germinate, first seen in part in Bismarck's social insurance and other legislation in Germany in the early 1880s or in the Liberal government's programme in Britain just before 1914.

Whatever the political repercussions, clearly the change from a liberal non-interventionist state to a more interventionist approach was to have significant geographical impact. As the state came to involve itself in every sphere of national life, its intervention made a greater impression on the landscape, directing and regulating in fiscal and planning policy the spatial pattern of the national territory, so welfare states have sought in particular to even out regional inequalities in economic and social well-being. While the corporatist approach — the state, management, and labour acting in concert — remains a common system to promote national efficiency, in the 1980s a feeling of a need to 'roll back the state' has marked several countries.

47

THE TOTALITARIAN STATE

Application of extreme political doctrines (whether to right or left) may change the role of the state form a liberal-democratic system with interventionist tendencies into a totalitarian system. The state comes to dominate dictatorially the life of the nation, regulating every aspect of society and the economy through the absolute command of an all-pervading state apparatus. Unlike the reformist collectivism that spawned the welfare state in Western Europe, the revolutionary collectivism of Marxist-Leninism has regimented everyday life into its own image in the Soviet Union, Eastern Europe, and, in modified form, in China, and several lesser countries. This pattern of the state also existed briefly in the fascist regimes interwar in Italy, Germany, and Franco's Spain, but is still present in some Latin American countries. Absolute *dirigisme* and dictatorship in highly centralized planned economies have a major geographical impact on the territorial pattern of the state. The current dilemma of states of liberal persuasion is to measure an acceptable level of interventionism against a fear of creeping totalitarianism.

In the liberal-democratic state capital has usually been an ally of the state apparatus but not under its control. Now a dilemma has appeared in the growth of gargantuan capital structures owing allegiance to no one state, immense transnational industrial and financial corporations able to shift the spatial emphasis of their activities rapidly from one part of the world to another and to 'play' the world money markets. These huge organizations have structural characteristics of a 'state' and it is perhaps not surprising that, irrespective of international relations between the west and the Socialist *bloc*, they have established a good rapport with state monopoly capitalism of the Soviet type (Levinson 1980). By their size and power, they have introduced a discordant note into the long-established relations in the liberal-democratic state between the state apparatus and capital, management, and labour, which have usually operated in a belief that their actions are to the *national* good, even when in dispute. The new transnational corporations have, however, posed their management and labour (and their investors) a dilemma, a need to balance their national common weal against that of their employing corporation. In this lies the germ of a threat to the accepted concept of the sovereignty of the state.

THE NATION-STATE AND THE FUTURE

We cannot asume that the nature of the state will not continue to change, for there are already developments suggesting possible directions, both for the individual state and for the global system. Between the sixteenth and the nineteenth century the old pattern of overlapping political jurisdictions and multiple competence, acceptable so long as states were small and shifting in extent and monarchs relatively weak, was replaced. There emerged the centralized state, claiming monopoly in the making of laws and their enforcement over a legally defined territory, independent of any outside authority. The geographical boundaries of the state and its competence and those of the nation it represented came increasingly to coincide: the nation came consequently to be the sole recognized source from which a state's legitimacy could be claimed. This concept spread from Europe to the other continents, though outside Europe, notably in Africa, newly independent states (organized in the received image of European experience as the great imperia disintegrated) have had in many instances to create their 'nation' from divergent and mutually hostile peoples, tribes, or groups entrapped by arbitrary former colonial boundaries. The rigours of this effort remain a major cause of instability.

The nation-state has shown itself to be economically, socially, and militarily successful, while as the political franchise has been widened, it has proved capable of satisfying the feeling of community in the nation, with support from competing and even conflicting social groups. The state apparatus, capital, management, and labour have found a common concern in the profession of acting for the national good. It has given a unified internal market as well as protection from outside for farming, industry, and commerce. Emotive feelings for nation and homeland have provided a popular support for the military as the guarantor of national security.

Nevertheless potential elements of weakness begin to emerge. Brunn (1984) has pointed to some of the new characteristics that may demand adjustments of the nation-state system, identifying ten spheres of change. The shrinking of the world has arisen through advances in transport and communication, so that distance has become less significant as a barrier to interaction, creating new linkages. International frontiers have become less meaningful as more problems become common to more than one political territory. Increasingly pressure groups have begun to represent global or macro-regional issues rather than local or uninational ones. A further field for adjustment has arisen from a shift in priorities from wealth generation by production to human and environmental issues of the 'compassionate society'. Citizens have been more and more faced by possibly conflicting demands between their transnational company employers, internationally organized professions, or other groups to which they may belong and their primary loyalty to their nation-state. Another factor has arisen through the widening of the horizon of the organization of space from the land surface on to adjacent seas, and the open ocean, and even into air space (including possibly into outer space itself) — how far should demarcation be along the lines of nation-states and how far for the 'common heritage of humankind'? A significant further factor has arisen from the growing ease of communication, tending to reduce the role of core areas and the centripetal–centrifugal forces between them and their peripheries: in place of this relationship there is a growing importance of being linked to the various communications networks, with particular importance for their nodes. The ranking of these nodes and the relations between them suggest potential for new governmental organization, but there is also the influence exerted by the emergence of new superstates or even commercial bodies of such power as to dominate the affairs of nation-states, creating a new distribution of power.

The unanswerable question is whether these adjustments can be made without recourse to war or civil commotion or to a return to an intolerable and stultifying degree of fragmentation that so long fettered the advance of European society in the middle ages. The greatest danger of fragmentation is that it makes domination and 'enslavement' so much easier for the stronger and larger powers. To satisfy all the myriad complexions of human society suggests a virtually limitless fragmentation into ever smaller independent political territories, among which at some scale there must come a threshold below which their viability ceases.

The most serious and active challenge to the nation-state comes as national markets have been transformed to a global scale as business and finance has shifted from a principally national to a transnational base, with immense corporations with a loyalty only to themselves cutting directly across the national dimension. Even the richest and largest national markets now give inadequate economies of scale. States, in a search for wealth generation and employment, vie with each other to attract the favour of such vast commercial organizations, often at considerable expense. Such is the mobility and power of these corporations that their attachment to any state lasts only as long as the life of a particular investment, posing a threat in many instances to the continuity of national policies. Most notable is the power of international financial bodies, which hold the key to the future of heavily indebted nations. Nevertheless the situation is a precarious one, for, if any serious debtor state were to default, the ensuing world financial crisis might become uncontrollable. Equally important are the internationally organized commodity markets, able to influence world trade, as seen in the generation of a worldwide recession through the sudden repricing of crude oil in 1973.

A further dilemma arises from the advent of nuclear weapons and their influence on military balance, since these pose a threat to the exclusive sovereignty of the state, making it impossible to continue the guarantee of security to its citizens. Fear of escalation to the nuclear threshold has doubtless constrained nations' readiness to resort to armed conflict. The spread of nuclear weaponry both vertically (the types according to flight range and complexity) and horizontally (among those states holding such weapons) has been conditioned by the efforts of the superpowers to match each other's armoury (a slow escalation) and to contain the number of countries possessing such weapons. Even states of considerable wealth face a dilemma of how to finance the rapidly rising costs of modern weapons systems, forcing them into ever closer military treaty systems and into a rising dependence on the patronage of superpowers wealthy enough to conduct research and development and to provide the most sophisticated weaponry. The first concrete measures to control armaments made by the superpowers in 1987 were as much the outcome of the rising economic burden of defence as that of popular pressures.

The third challenge arises from the political and social atmosphere becoming steadily more conducive to minority groups asserting their self-determination and challenging the sovereignty of the nation-state over them. This is a particular embarrassment, for they seek the same principle of self-government on which the legitimacy of the state's sovereignty was itself founded. Whereas the economic and military challenges are externally generated, the pressures from minorities are more difficult since they are internal questions. Clearly some compromise between the centralist, assimilative, or even repressive structure of the nation-state and the centrifugal, fragmenting, and divisive claims of minority groups needs to be sought. An approach to this problem downwards may be seen in the abortive devolution debate in Britain in the 1970s or, on a wider scale, in a surrender of some measure of sovereignty upwards by membership of the European Community.

Beetham (1984) has suggested three ways forward for the nation-state. First, an ill choice and one preferably never made, to seek to reassert its exclusive sovereignty against forces both internal and external. This could precipitate its collapse into anarchy. A second, to seek plurality of political structures, of which the nation–state could simply be one, among sub–national, macro–regional, and global ones. For this the necessary popular social and political impetus appears lacking. The United Nations, a move in this direction, has struggled to gain appropriate recognition and even a toehold over its members' sovereignty without great success. Third, most possible, is a modest relaxation of the state's exclusive claim to sovereignty, though experience in Western Europe suggests that building supranationalism is a slow accumulative process, not always without reluctance through resentment or fear.

3

Territory

Human societies have a strong attachment to territory, seeking to exercise absolute control over tracts of country regarded as vital living space. Hunting and nomadic groups have recognized territories, but definition becomes really important for sedentary agriculturalists, so land division was recorded as early as Sumerian times. The Romans carefully demarcated their imperial territories, private estates, and even parcels of land. The growth of a monetary economy in the middle ages made legal definition of land parcels and titles increasingly important as land ownership became an ever more attractive form of individual wealth and a lucrative taxable asset. This encouraged states themselves to define more carefully the country over which they claimed sovereignty, so vaguely defined marches were replaced by exactly demarcated frontier boundary lines. In modern times the whole land surface of the earth has been divided into legally defined territories and the process has begun to be extended over the sea and ocean floor.

Quite apart from its value as living space, society sentimentally treasures its territory as one of its most sacrosanct possessions. The national territory usually has a prominent place in the nation's iconography, with the homeland personified as the 'fatherland' or 'motherland' and attachment to it expressed in poetry and song, with the surrender of any of it regarded as unacceptable. Nevertheless feudalism, built of a hierarchy of loyalties, had regarded territory less importantly than the modern nation-state, so monarchs laid less value on compact territories than was later the case. The absolutist state became more dependent on territory as a source of wealth, and consequently more interested in a careful definition of the lands it held, while the shift to the state idea centred around the nation intensified the meaning of territory. Nations have usually clearly perceived ideas of the extent to what they regard as rightly their homeland. Unfortunately, not infrequently, the territory regarded by one group as rightly its own overlaps a similar perception by another group, creating conflicting claims to that particular tract of country.

CLAIMS TO TERRITORY

Claims to territory usually have a strong historical bias, emphasizing it particularly as a part of the claimants' homeland. It is, however, not unusual for claims to be tenuous, based on the claimants' own interpretation of history. As an example, a Polish nationalist organization between the wars published a propaganda postcard claiming territory on Poland's western borders, including Bohemia, on the grounds that it was united with the Polish Crown in the years AD 1003–4! A recent example has been the divergent interpretations of legal and historical ties made by Britain and Argentina over the sovereignty of the Falkland Islands and it is perhaps not a coincidence that this dispute has surfaced as the belief has grown that the offshore waters may be oil-bearing. Claims made on legal or historical arguments often conceal strategic or economic undertones, like Soviet claims in 1946 to territory along its European borders; they lacked historical conviction but had sound strategic reasons.

For long-term advantage states usually seek to turn *de facto* possession into a *de jure* holding by establishing their entitlement under international law through recognition by the community of nations. For long-established national territories there is usually tacit *de jure* recognition of sovereignty, a situation covering the greater part of Europe. Nevertheless formal legal documents, such as those of the Congress of Vienna of 1815 or of Versailles and its related deliberations of 1919–20 exist, notably for the so-called Succession States of the Habsburg Empire.

The expansion of Eurpean powers overseas from the seventeenth century onwards generated another form of legal recognition, the demonstration of 'effective occupation' of territory. What it really meant was disputed in many conflicting claims, but it was certainly the means by which sovereignty was extended in North America, South America, Siberia, and Africa. Annexation had to demonstrate such aspects as extension of the occupying power's legal system to the new territory, as well as providing it with effective administration and some form of representation in the claimant's legislature.

One of the most common ways of changing the ownership of

territory has been military conquest, though this in itself has not been accepted as a basis for a *de jure* claim to sovereignty. Such recognition has usually been through the subsequent formal annexation and the demonstration of 'effective occupation', confirmed in formally concluded documents of cession or a peace treaty. To establish a *de jure* claim, the former owner must have renounced his claim or *de facto* abandoned reconquest. Declarations of annexation while conquest was continuing have never been considered appropriate legal ground, as in the Italian annexation of Tripolitania in 1911–12 and of Abyssinia in the 1930s. Sometimes a simple act of occupying territory and raising the flag may be sufficient base for a claim of sovereignty — Britain has sought to establish definitive ownership of Rockall by landing a party of Royal Marines and raising the Union flag, though this has been questioned by Iceland and Ireland.

Where territory claimed by one state has been effectively occupied by another over a period of time, the latter may seek to establish *de jure* sovereignty by 'prescription', so long as there has been no serious and regular objection from the original owner to the occupation. Though now infrequently invoked, cases do still occur. Regular declarations by the original owner of his claim to continued sovereignty helps to prevent the territory passing to the occupier by default. The International Court of Justice arbitrated in favour of Sweden in its dispute with Norway over the Grisbardana district, where Sweden had for a long time exercised *bona fide* sovereignty without objection. The same court upheld Denmark's claim to Greenland in that it had exercised territorial sovereignty since the seventeenth century and had entered into international agreements over the territory without resistance or objection from other states. The same argument upheld British possession of the Miquiers and Erechos Islands against France. Over a long period Britain had exercised territorial rights without French objections and France had not laid claim to the islands until long after the original British occupation.

Acceptance of the concept of self-determination following the First World War led to the development in the 1930s of a legal view of the unacceptability of annexation as a *de jure* claim to territory, noted in the Briand-Kellogg Pact of 1928 and subsequently reaffirmed. The so-called Stimson Doctrine made annexation by violence unjust in the eyes of the community of nations and has been used as a basis for legal argument. Annexation was rejected in the Atlantic Charter of 1941 and its unacceptability written into the United Nations Charter. Assimilation of territory effectively amounting to annexation may be achieved through duress or coercion falling just short of military conquests. Such tactics have on occasion found reluctant acceptance among the community of nations. Examples include the Japanese absorption of Korea in 1910 and the German *Anschluss* with Austria in 1938. The Soviet 'assimilation' of the Baltic republics in 1940 has been recognized *de facto* by Britain but not at all by the United States. An unusual situation arose in 1938 when Britain and France put pressure on Czechoslovakia to surrender its 'Sudeten' German districts to the Third *Reich*.

In medieval Europe transfer of territory without dispute was common, but it must be seen in the extreme complexity of feudal relationships and the personal bonds of fief, homage, and vassalage which were often vacillatory and shifting. Pounds (1951) points to the interpenetration of feudal obligations, exemplified by the Counts of Champagne who were vassals of the Hohenstaufen Emperor for three of their French territories, over which the French king could exercise no feudal rights, though remaining their king. A careful study by Shaw (1986) has pointed to the Treaty of Westphalia of 1648 as marking the crucial point in centring international law upon defined territorial units. Peaceful transfer of territory has also arisen through purchase, and the United States of America was built substantially through this means — most notably the vast Louisiana Purchase from France, a bargain at $15 million, even in 1803; by the Gadsen Purchase from Mexico in 1853; and Alaska, bought from Russia in 1867 for $7.2 million. Other forms of transfer have been several small colonial territories from Britain to now independent Dominions (e.g. Cocos Island in 1955 and Christmas Island in 1959 to Australia); the French territories of Pondicherry transferred to India by a Treaty of Cession in 1956; and the Spanish agreement to cede their Western Saharan territory conjointly to Mauretania and Morocco in 1975. This was not well received by several countries since no consultation with the inhabitants took place, while in 1979 Mauretania withdrew from its share of the territory, then occupied by Morocco. Transfer of territory often arises from agreed frontier rectifications, as in the case of considerable revision of the Jordan–Saudi Arabian boundary in 1965, the most important outcome being the strengthening of the territorial position of Jordan's port of Aqaba. Under the Quadripartite Agreement on Berlin of 1971 an exchange of territory between West Berlin and the German Democratic Republic took place, when six exclaves and a small piece of border territory (15.5 ha) were given to the Republic in exchange for two exclaves and two adjacent border areas (17.1 ha) and, to offset the difference in area, a compensatory payment was also made to the Republic. The principal aim was to ease access from West Berlin to the small settlement of Steinstücken and the Eiskeller district.

Conflicting claims to territory have sometimes been defused by establishing neutral zones, administered conjointly or by a third party. An agreement in 1981 ended the sixty-year existence of the Saudi-Arabian–Iraq neutral zone by equal division between them. Between 1816 and 1919 Moresnet near Aachen was a neutral and customs-free territory, jointly administered by Prussia and the Netherlands (1816–31) and then Belgium (after 1831). Similar arrangements may be made in 'international territories', as in the Franco-Spanish Treaty of 1912 internationalizing and demilitarizing Tangier, which since 1956 has been a free port, though part of independent Morocco. After 1920 Danzig was made a free city under the League of Nations, because its completely German character made it inexpedient to incorporate it into Poland which was allowed extra-territorial rights in the port. Joint territorial administration is also found, as in Vanuatu (New Hebrides) before independence, operated since 1906 as an Anglo-French condominium, with some powers joint but others unilateral. Andorra is a longer-standing condominium, administered by France and the Spanish Bishop of Urgel.

Some territoral arrangements may be seen as 'servitudes', constraints on the sovereignty of a state in its own territory, representing an obligation to permit certain action there by others. The Act of Mannheim of 1868 imposed an obligation on all riparian states on the internationalized Rhine to allow free navigation of vessels under their flags. As the result of boundary changes in 1919, the Czech town of Frýdek-Místek had its water supply partly in Poland, so a clause was inserted in the treaty defining the new boundaries allowing Czech waterworkers free and unimpeded access at any time to the relevant installations in Poland. The Treaty of Paris of 1920 imposed an obligation on Norway to allow other countries to conduct commercial operations (notably coal-mining) under their own laws on Spitsbergen (Svalbard), though only the Soviet Union has taken advantage of the concession.

A further special territorial situation is when a state concedes special rights to another on its territory without loss of title or sovereignty, often through 'concessions' or 'leases'. Military bases commonly fall in this category: examples are the United States' base at Guantanámo in Cuba, the British sovereign bases in Cyprus (an extreme example), or the Soviet Base 'leased' from Albania on barren Saseno Island from 1945 to 1961. In 1963 the Soviet Union made a fifty-year lease to Finland of a three-kilometre-wide strip along the important Finnish export route by the Saimaa Canal giving access to the Gulf of Finland at Soviet Vyborg (Vipurrii). Last century China was forced to make concessions to Britain, Russia, France, and Germany in its so-called 'treaty ports': after 1919 the Russian and German leases passed to Japan without even Chinese consultation. An interesting example of leasing with complete sovereignty was the British lease from China of the 'New Territories' of Hong Kong for ninety-nine years in 1898, strengthening the viability of the small Hong Kong Island ceded to Britain by China under the Treaty of Nanking of 1842 and the Convention of Peking of 1860. Under an agreement, concluded in 1984, China will recover all this territory when the lease expires in 1997, but has undertaken not to change the economic and social regime in it for the subsequent fifty years.

THE TERRITORIAL STRUCTURE OF THE STATE

Every state territory is in a sense unique, by virtue of its size, shape and geographical character, yet every territory has in common similar spatial elements in its structure. Fig. 3.1 is an attempt to construct a model of these elements, though their definition and delimitation in reality are not always easy. Perhaps the most important element in the territory is the focal area from which the major political, economic, and social impulses come, usually containing the national capital. It is not uncommon that this focal area encompasses the historical nucleus or core where national identity first germinated and the state idea, its *raison d'être*, was formulated. The second major element is what we may term the total state territory, held *de jure* or *de facto*, its outer limit marked by a formally delineated frontier with neighbouring countries. We may also discern a third element, the idea of the effective state territory within the total territory, more prominent in the past, when transport and communications from the focal area were less certain and made the task of government in outlying districts more difficult. Remoteness or physical inaccessibility, even the pressure of dissident or separatist movements in particular regions, weakens the control of central government in the peripheries. A fourth element may occur where the territory controlled by the state does not encompass all the settlement area of the nation — where compatriots live under foreign rule, the state may claim that territory as an *irredenta* (from Italian, 'unredeeemed').

The nature of the state may affect the expression of these elements within its territory. We would expect a weak state to have a poorly

Figure 3.1 Structural components of national territory

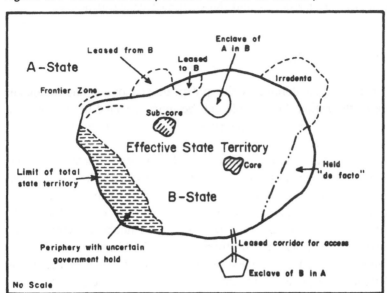

developed focal area and its effective state territory to be quite limited, with government hold tenuous on the periphery, the sort of situation found currently in Central American republics like San Salvador and Nicaragua. On the other hand, a strong state, like France, may be expected to have a vigorous focal area, clearly directing its management, just as little if any of its territory will be beyond the effective rule of central government.

The ease with which central government may exert its will over its territory and the effectiveness of that control over different parts of the country will be influenced by several geographical factors. Considerably different management challenges arise in a vast territory with great distances (like the Soviet Union), even with modern communications technology, compared with those in a small-to medium-sized country, like Poland or Belgium, though factors such as density of settlement and economic potential are also critical. The tasks of government are also affected by the shape of the national territory and by its relief through their influence on accessibility and on regional interrelationships. External factors arising from the relative global location of the territory may also influence a government's internal management problems. Geographical location, defined by latitude and longitude, is fixed, but geopolitical location, of special importance to the state's fortunes, will vary with changing relations with adjacent territories and the shifting balance of power.

THE FOCAL AREA

The focal area dominates the political, economic, and financial life of the nation, but its role can be much affected by its location within the geometry of the state territory, affecting the significant core–periphery relations. In the United Kingdom, Southeast England centred on London is the focal area, pre-eminent in governmental and financial affairs, but it has strengthened its commercial and industrial position since early this century at the expense first of the North of England and Scotland, and later the English Midlands. Its extreme southeast position is no doubt a factor in the tension existing with the northern part of the kingdom which feels remote from and neglected by central government. Whereas in Britain some government functions have been delegated to Scotland and to Wales, Paris has remained the focal area of a more highly centralized state for much longer, though historically this role has been weakened by its position close to the northern periphery and vulnerable to enemy incursions. History has left some countries with focal areas of a magnitude disproportionate to their contemporary needs, as in Vienna and Budapest. Vienna has sought an international role to offset its loss of a major empire to serve, while Budapest has maintained a powerful and continuing attraction power to industry so that about a fifth of the total national population now lives there. The proportion of total national population living in the focal area is not, however, always a measure of the strength of its influence over national life. In the Soviet Union little more than a twentieth of the total population lives in the focal area around Moscow, but by the nature of Soviet bureaucracy the influence of the Moscow region is extremely great, despite the organization of the Soviet Union as a federation. This does not imply that federations do not have focal areas: the United States has a clear focal area, embodied in the great urban agglomeration of Gottman's 'Megalopolis' embodying the main centres dominating the political life of the nation: Boston, Philadelphia, and Washington, DC, as well as the unchallenged financial centre, New York.

THE CAPITAL CITY

Within the focal area is commonly the capital city (so often also the nation's primate city) which attracts particular prestige and represents effectively the 'control panel' from which the state apparatus is manipulated, for it is normally the seat of the legislature, judiciary, the ministries, and government departments around which diplomatic missions cluster. The capital occupies a key place in the nation's iconography and has often been developed lavishly at the expense of the rest of the country. In some countries, mostly on historical grounds, the capital functions are divided between two or more cities. In South Africa Pretoria, focus of Boer aspirations, is the government seat and capital, but Cape Town, with strong British traditions, is seat of the legislature. In Libya Tripoli and Benghazi are regarded as equal in capital status.

Location of the capital within the geometry of the state territory depends on factors of population distribution, economic development and especially on historical experience. Though Warsaw and Madrid appear to function well from their central location, Alice Springs would be hardly a suitable site in Australia! In federations the capital is often given the semblance of being independent of the federal members by occupying its own territory, as Canberra in the Australian Capital Territory or Washington in the District of Columbia in the United States. The capital may also be located to emphasize a particular policy, for example, Brasilia, the new federal capital for Brazil, sited to point to the undeveloped interior and to reduce the dominance of São Paulo and Rio de Janeiro. Peter the Great built St Petersburg on the border of his empire as a 'window on the world', open to modernizing influence and away from the conservatism and traditionalism of Moscow. A strong symbolism was embedded in the move back to the more central site of Moscow after the Revolution, but it also emphasized the dominance of the Great Russians in the Union, for the city is the historic focus of Great Russian national identity. Symbolism also underlay the move in 1920 of the Turkish capital from Istanbul to Ankara, while likewise the British in India chose to move the Viceroy's seat from Calcutta to the magnificent New Delhi, for Delhi had been capital of the seventeenth-century Mogul Empire.

The problems of selecting a capital are well illustrated by the search for a 'provisional' capital, on formation of the German Federal Republic in 1949, to replace Berlin, when several cities by tradition, size, and morphology were worthy of consideration. Two major cities, the Hanseatic port of Hamburg and the former Bavarian capital,

61

Munich, while possessing suitable attributes, were considered too peripheral in north and south. Too close proximity to the new inter-German border disqualified the old royal seat of Hannover, while this factor, added to inaccessibility from the Rhine basin, ruled out Kassel. Despite its good location and fortunate historical associations, Frankfurt was unacceptable because it was the main American head-quarters, while Wiesbaden was excluded as it was too close to Frankfurt. With Cologne heavily damaged, the choice fell on the relatively small university town of Bonn, which had long associations with Rhenish liberalism, and was favoured for personal reasons by the Federal President, while it also lay on the Rhine, foreseen as the major economic axis of the new republic.

Location of the capital has sometimes been a compromise between conflicting regional interests. Washington, DC, was located between the contrasting cultural and economic spheres of the late eighteenth-century *states* of the Union, while since 1867 Ottawa has served as capital for the Canadian provinces, lying intermediately between English- and French-speaking areas. Before that time the Canadian parliament had alternated its sittings between English-speaking Toronto and Francophone Québec. Of course, location is not immutable, as already noted, and a shift in location may be a response to spatial shifts in the structure of the state. In response to the opening of Japan to the world in the 1860s, the capital was transferred to the port of Edo (Tokyo) from the traditional interior capital at Kyoto. In newly independent countries the colonial capital commonly does not meet the new national needs, so Malawi has moved its capital from Zomba to the more central Lilongwe; Tanzania's seat of govern-ment has gone from coastal Dar-es-Salaam to interior Dodoma; while Nigeria has been building a new federal capital at Abuja, well inland from colonial Lagos. The building of Gaberone in Botswana aimed to have the capital on its own territory, as previously the administra-tion had been in Mafeking, the railhead in South African territory. Capitals of this kind are sometimes termed 'introduced' in contrast to 'permanent' capitals of long historical standing like Paris or London.

The older tendency to concentrate as much as possible of central government in the capital has slackened as modern communications systems have made decentralized location of day-to-day operations of government departments possible. Decentralization, begun for strategic reasons interwar, has been a means of bringing employment to disadvantaged areas, so, for example, in Britain large social-security head offices are located in Newcastle-upon-Tyne, vehicle licensing in Swansea, and television licensing in Bristol, among others. Federal offices have also been diffused in West Germany, with ministerial branches in West Berlin, Frankfurt, and Nuremberg, as well as the Federal Constitutional Court in Karlsruhe.

THE CORE AREA — THE HISTORICAL ORIGIN OF THE NATION

We have already examined the role of the 'focal area', the focus of political, economic, and social forces, within the contemporary state territory, but we may also distinguish an important historical component, termed the 'core area' by Whittlesey, expanding on a concept proposed by Ratzel. The conceptual dilemmas have been admirably summarized by Muir (1975), especially as some students have sought to define the 'core area' on non-historical criteria. The term should be used only for that part of the national territory where the historical idea of the nation germinated. The historic core will commonly but not invariably lie within the contemporary focal area, though it may not be easy or even possible to identify such an historical area for some nations, despite their having a readily apparent modern focal area. Pounds and Ball (1964) made a thorough investigation of core areas and their development in the countries of Europe, describing their essential geographical characteristics (Fig. 3.2). Among important common criteria, the core area had to be easily defended and able to generate sufficient wealth to pay for its defence and to play its proper role in the power politics of the time. For the development of most core areas, this implied fertile country, well cultivated, and adequately productive in the measure of the age, as well as a population sufficient to maximize the advantages from local resources, besides a vigorous long-distance commerce to supply materials not available locally. Pounds and Ball also examined whether the expansion of the nation from the core area was seen as a grand design, with a clear concept of the future geographical limits of the state, and a conscious effort to reach them. Pounds, with his customary good historical sense, rejected ideas such as that of Sorel in nineteenth-century France that the early French kings conceived of a kingdom bounded by 'historic' or 'natural' frontiers.

The geographical characteristics of the core area defined by Pounds and Ball (1964) may be found in several different parts of the national territory, but in one of these parts a significant human factor triggered off germination of the nation. The vigour and aggressiveness of the Capet family from their estates on the Seine brought Hugh

Figure 3.2 **The core areas of European nations**

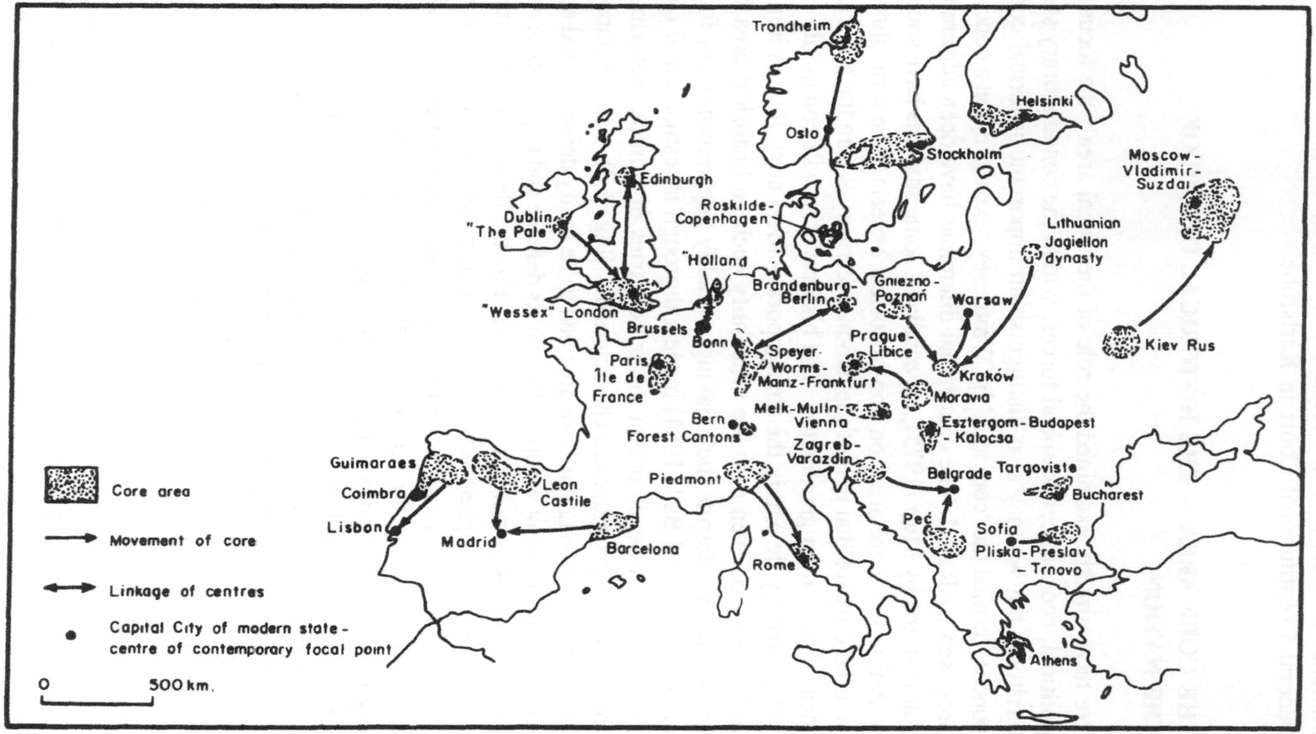

Capet in AD 987 the then rather empty honour of election as King of France, on which the family built its leadership of the nobility of the French lands. Similarly the rulers of the Moscow principality, in leading the Russian princes against the threat from marauders from the steppe, used their position to expand the limits of their own territories. Though in neither case does there seem to have been a 'grand design', perhaps 'the inevitable logic of geography lying at the root of all history' coloured the direction of a conscious, if unplanned, expansion.

The position of the core may shift as the process of nation-building proceeds. The first signs of an emergent Polish nation was in the tenth century AD at Gniezno, but by the fourteenth century the focus of nation-building had moved to Kraków, while from the late sixteenth century Warsaw became the centre around which the modern nation crystallized. Some nations which were formed quickly or with weak identity may display no clear core. Belgian 'nationality' emerged in the Brussels riots of 1831 that saw it independent of the Netherlands, yet Brussels has since become a point of weakness and friction between Walloon and Fleming. It is difficult to describe a core area of German national identity. In Frankish times there seemed to be a core emerging in the middle and upper Rhineland, but the failure of the Holy Roman Emperors to exert an effective authority made this abortive. When national feeling again awakened in the nineteenth century, the Rhineland again played a significant but abortive role. The final coalescence of the Germans into a political nation came through Prussian pressure from Berlin. We may also consider the emergence of the American nation centred around the common sentiments of the original thirteen colonies on democratic government. Much of the initial pressure came from the New England colonies, but they were supported by others, notably Virginia, New York, and, in particular, Pennsylvania. If we seek one place that played a crucial role, then the choice must fall on Philadelphia where conceptual principles for government to serve an 'American way of life' were formulated.

STATE TERRITORY — MANAGEMENT DILEMMAS

The political map is a mosaic of the total state territories of the countries that comprise it, although few, if any, maps ever distinguish between territory held *de jure* or *de facto*. Each territory is defined from its neighbours by a finite boundary line, though there are occasional uncertainties even in modern times. Effective rule by central government from the national focal area may not, however, extend over the total state territory. Consequently we may find central government has full control only over a smaller area, the *effective state territory*. This may arise from different reasons, but commonly there is a factor of remoteness or physical inaccessibility making full control difficult — the Ottoman Turks never managed to extend their full authority over the remoter, more difficult mountain country of the Balkans. Dissidence and disaffection may be another factor, as in the Kurdish districts of Iraq and Iran. In Central America, in El Salvador and Nicaragua, political insurgency, fuelled from across the frontier, has considerably reduced the effective state territory in both countries, while guerrillas control large southern tracts of Morocco, effectively pushing back central government influence. The difficulties are usually greatest in fragmented archipelagic state territories.

In the middle ages the difference between the notional total state territory and the area effectively controlled by the king was commonly wide — there is plentiful evidence of the limited range of effective royal authority in Scotland, while the Capet kings of France for long exercised real control over little more than their own estates of the royal domain.

Tightening the hold of central government upon its peripheral territory has usually been the aim of regional policies. Helping backward or disadvantaged areas to improve their economy and infrastructure invariably extends the power of central government. Regional policy often involves an element of 'internal colonization', as in Siberia, where Slavonic settlement, vital for its economic development, aimed strongly at stamping this vast territory as essentially 'Russian land'. The seventeenth-century 'Plantation' of Ulster by Scots settlers certainly aimed at consolidating English domination of Ireland, leaving a bitter legacy for the twentieth century. At present the Chinese are pursuing a similar policy of military colonization to tighten their grasp on their western periphery in Turkestan. Central government control over the periphery has been made easier by the erosion of distance with improvements in transport and communication systems but has increased the pressure for standardization and uniformity. In eighteenth-century Siberia couriers and officials could take months to reach their destination from St Petersburg, so provincial governors had to be virtually independent, whereas today passage of information from Moscow is almost instantaneous and air travel measured in hours.

The relationship of total territory to effective territory is a key element in the tensions of a social, political, or economic character between core and periphery. As the focal area tends to concentrate wealth and power upon itself, the periphery feels neglected, misunderstood, or patently ignored, with consequent disenchantment. In such a scenario, unless there is strong national coherence, the periphery may become secessionist. State policies have increasingly sought to distribute economic development and social conditions more evenly over the territory, though there has usually been a reluctance to devolve political power, perhaps in fear of generating federalist tendencies. In essence the coherence of the state territory is assured by manipulation to balance centripetal against centrifugal forces, with serious instabilities generated if that is not achieved.

The countries of the world show a remarkable range both in geographical area and in population magnitude (Table 3.1). Even with modern technology, government of very large territories poses difficulties for national coherence, integration, and administration,

Table 3.1 The world's largest and smallest states

Largest

By Area (million km^2)			By population (million)		
1	Soviet Union	22.402	1	China	1,015.1
2	Canada	9.970	2	India	683.8
3	China	9.597	3	Soviet Union	276.3
4	USA	9.167	4	USA	234.0
5	Brazil	8.512	5	Indonesia	147.5
6	Australia	7.682	6	Brazil	132.6
7	India	3.167	7	Japan	119.5
8	Argentina	2.778	8	Bangladesh	89.9
9	Sudan	2.506	9	Pakistan	88.0
10	Algeria	2.382	10	Nigeria	82.4

Smallest

By area (km^2)			By population (thousands)		
10	Grenada	344	10	Seychelles	64.7
9	Maldives	298	9	Kiribati	60.0
8	St. Christopher	261	8	St. Christopher	44.1
7	Malta	246	7	Andorra	41.0
6	Liechtenstein	160	6	Monaco	27.0
5	San Marino	61	5	Liechtenstein	26.5
4	Tuvalu	24	4	San Marino	21.6
3	Nauru	21	3	Nauru	8.4
2	Monaco	1.95	2	Tuvalu	7.3
1	Vatican City	0.44	1	Vatican City	1.0

Source: Stateman's Yearbook, 1985–1986

so it is perhaps significant that the world's federal states are particularly well represented among the ten largest countries by area and by population. The difficulties are intensified if there is much awkward terrain or extensive sparsely settled tracts. Consider the administrative dilemmas of the Soviet Union, whose huge longitudinal extent means Moscow goes to bed as Vladivostok gets up! Mainland China, though only half the area of the Soviet Union, poses problems for government by having to manage almost a quarter of mankind, with variation in density between some of the world's most densely settled farming country in the Yangtse delta, and the empty deserts of the western interior, or the thinly peopled mountains of Tibet. Many small countries face problems of management arising from their limited extent and small population, despite being relatively wealthy in *per capita* terms. Effective defence is always a major dilemma, while the smallest of all may find it hard to maintain all the functions of nationhood. Small countries commonly seek neutrality and ensure it by providing an 'unbiased' base for international meetings or organizations (Switzerland, Luxemburg), while they may also fulfil special roles, making them indispensable to larger neighbours (Swiss banking and financial services, Dutch maritime commercial services). The tiniest countries usually rely heavily on larger friendly neighbours for many social and economic services, as in the case of Monaco *vis-à-vis* France or Liechtenstein's economic and customs ties with Switzerland.

Problems of management for government may also arise from the shape of the state territory and its topography (Fig. 3.3). Easiest is a compact territory, especially with some semblance of a circular or hexagonal form and without serious interior physical barriers. Poland and Hungary provide useful examples. Though compact, France long faced a defence dilemma on its northeast 'salient' gripped in a German territorial 'pincer', while the rough and often poor country of the Massif Central canalized movement around its flanks. The neatly compact shape of Rumania is marred by the serious barrier of the huge bow of the Carpathians sweeping across the interior through which there are few good routes. The country to the west has a strong Pannonian orientation, that to the east a distinctly Euxine orientation. The rectangular shape of Yugoslavia is cut lengthwise by the inhospitable Dinaric mountains, so their western, Adriatic face looks to the Mediterranean Basin, while their eastern flanks are distinctly Pannonian.

Management of countries whose length is much greater than their width can be most awkward, especially if the focal area and capital lie towards an extremity. The excessive elongation of Chile makes

Figure 3.3 **The shape of national territories**

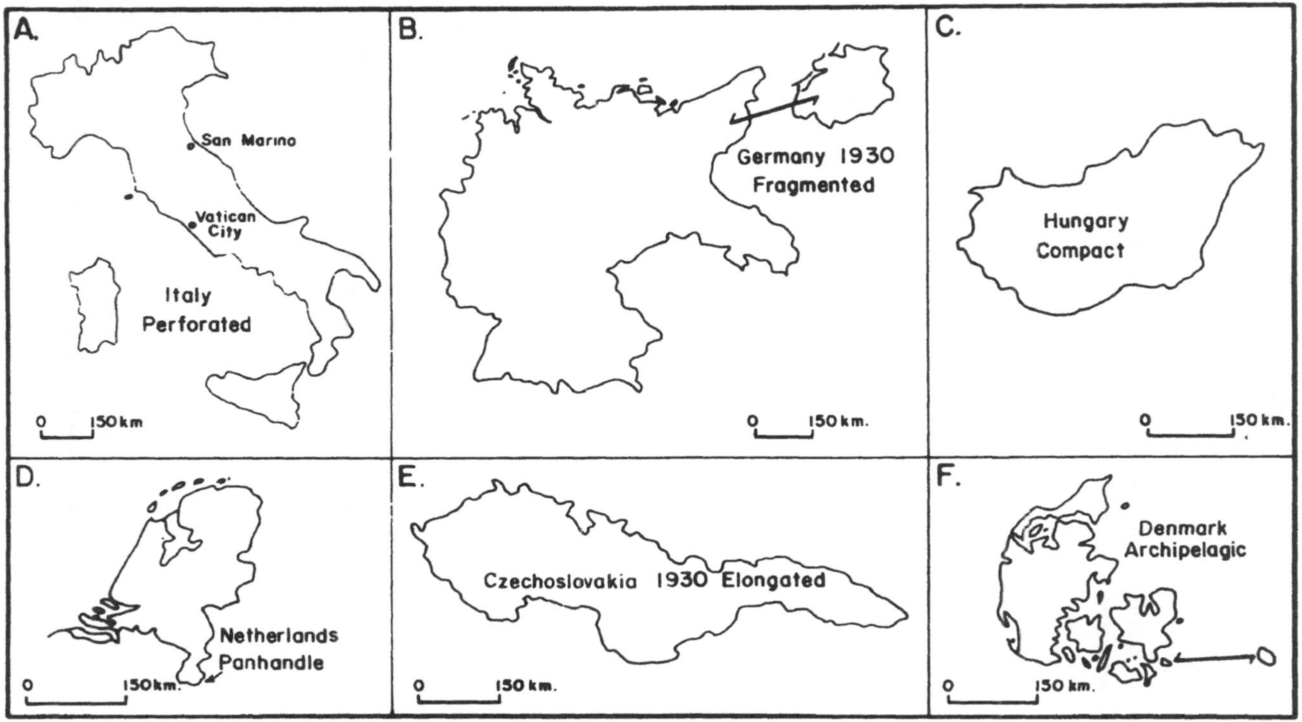

A.
San Marino
Vatican City
Italy
Perforated
0   150 km

B.
Germany 1930
Fragmented
0   150 km.

C.
Hungary
Compact
0   150 km

D.
Netherlands
Panhandle
0   150 km.

E.
Czechoslovakia 1930 Elongated
0   150 km.

F.
Denmark
Archipelagic
0   150 km.

the extreme northern desert and mountainous fjord country of the extreme south remote from the central focal area around Valparaiso. Sandwiched between the Andes and the Ocean, the country is some twenty times longer than it is broad. We may see something of this problem reflected in Britain, with the focal area around London perceptually remote from the empty and inaccessible mountain and island world of northern Scotland. In Czechoslovakia shape has also played a role in accentuating economic, social, and political contrasts between Bohemia-Moravia with the focal area of Prague and 'distant' Slovakia.

Particularly awkward problems can arise where there are long territorial panhandles or narrow corridors or waists. Often remote and inaccessible, these are difficult to defend and cramped in economic potential. A good example is the Limburg panhandle around Maastricht in the Netherlands, whose economic welfare has benefited considerably from the greater mobility across international borders generated by the European Community. Panhandles have sometimes been created for distinct geopolitical reasons: the Wakhan Strip in eastern Afghanistan was designed to keep the Russian and British empires from direct territorial contact, while the Caprivi Strip in German Southwest Africa was to give Germany access to a frontage on the potential trade artery of the Zambesi. Some panhandles are effectively corridors to give access to the sea (e.g. the former Polish and Petsamo corridors or Zaire's access to the Congo mouth).

Difficult political relationships arise from the fortunately rare 'perforated' territory, where one state territory surrounds totally or almost totally the territory of another sovereign state. South Africa totally surrounds the independent country of Lesotho and very nearly encloses independent Swaziland, and Italy totally encloses both San Marino and the Vatican City (created under the Lateran Treaty of 1929). Equally awkward for government are fragmented territories, where the state territory is broken into two or more pieces, a not uncommon situation in Europe until the nineteenth century. The situation arose for Germany between 1919 and 1939, when East Prussia was separated from the main territory of the *Reich* by the Polish Corridor (likewise awkward for Poland). Fragmentation can be a recipe for disintegration where national coherence is weak, as witnessed by Pakistan in 1971 on establishment of an independent Bangladesh. Particularly inconvenient to administer is the archipelagic state territory, reflected in the political history of Indonesia, Malaysia, and the Philippines.

Enclaves were common in medieval Europe, but Robinson (1959) raised the important question of why such apparent anachronisms survived so long and even new ones were created in modern times (Fig. 3.4). This is not just in Europe, for East and Prescott (1975) have pointed to the persistence since the seventeenth century of numerous examples in the Indian subcontinent. Nevertheless most enclaves are small and a minor irritation to the parties involved. In Europe what Robinson terms 'normal enclaves' still exist in four instances, all historical survivals, each under 25 km² and with less than 1,500 people. The remarkable collection of land parcels forming the Belgian enclave in the Netherlands at Baarle Nassau and Baarle Hertog arises from peace-making in 1648 confirming a confusing feudal situation. The German enclave in Switzerland at Büsingen remains from a deal in 1723 when the Habsburg sold the suzerainty of the Hegau to Schaffhausen but isolated Büsingen by withholding it out of spite for its attitude to them some thirty years earlier. Campione remained in Italy, because, when the Swiss conquered Ticino, they did not take it out of respect for a Milanese monastery that held the fief. Llivia remained Spanish through careless definition in the Treaty of the Pyrenees (1659) which allocated villages in the Cerdagne to France but overlooked the fact that Llivia was legally a town and thus excluded.

Parts of the national territory reached only practically across the territory of another state were termed by Robinson 'pene-enclaves'. Typically found in mountain country, proper road access in most instances could be provided from the home state at a prohibitive cost, although sometimes such provision is impossible, as at Jungholz in the Tirol, linked to the rest of Austria only by a 1,000-metre-wide strip of territory across high mountains. Pene-enclaves often come to use primarily the currency and postal services of the state into which they protrude and may even be within its customs zone. As 'virtual enclaves' Robinson notes the presence of lands and buildings in and around Rome declared extra-territorial to the Kingdom of Italy under the Lateran Treaty of 1929.

Enclaves survive through the interaction of three forces — their sovereign state, usually reluctant to surrender them; the state in the middle of whose territory they lie, generally ready to welcome their integration; and the enclaves themselves, normally seeking the conditions which would benefit them most. Whereas India forcibly occupied French and Portuguese enclaves, Switzerland has consistently sought, without much success, to absorb its enclaves by peaceful agreement, even exploring after the Second World War possibilities of exchange of territory with Germany to eliminate Büsingen and the small Varena Hof estate, though suitable compensatory Swiss territory could not be found.

Figure 3.4 Exclaves-enclaves.

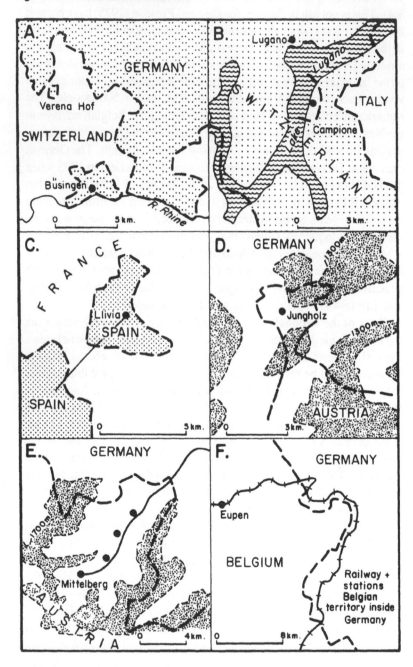

Adapted in part from Robinson, G.W.S., *Exclaves*, Annals of Association of American Geographers, vol. XLIX, 1959

Enclaves can be a source of international friction. Examples have been Swiss disapproval of Italians opening a casino in Campione, while free access to Llivia was overcome by a designated neutral road link from Spain. Access by land and air to West Berlin remains a matter of dispute between the Soviet Union and the Western powers. Enclaves may be the seat of potentially unfriendly actions: West Berlin is a base for broadcasts likely to influence public opinion in the East *bloc*, while East Berlin has used West Berlin as a means of penetrating the west for intelligence operations, as well as being a route for refugees moving from the East *bloc* into Western Europe.

4

The frontier

Perhaps no aspect of political geography has attracted more interest than the frontier, the critical interface between states, and a special element of the territorial question. The frontier is neither romantic nor exciting but a brutal fact, a constant potential cause of friction with neighbours. The most difficult task in international relations is to arrange frontiers that satisfy the beliefs, needs and honour of the parties on each side. Sadly changes in frontiers have seldom been achieved peacefully: change has mostly been attained through war, while war has almost inevitably changed frontiers between belligerents, even if they were not the primary cause of grievance. Essentially a zonal concept, a belt of transition between one state and another, the frontier is seen as a danger zone liable to enemy incursions, often perceived as better avoided, so it has usually been marked by economic and social differences from the rest of the country.

Early kingdoms lay separated by broad no man's lands in inhospitable terrain. As settlement spread, and especially as a growing monetary economy made land and people taxable assets, more careful and exact legal definition of ownership was needed, while emergence of national identity gave the frontier an added significance in national iconography. Formal legal ownership of all land made carefully defined linear boundaries a prerequisite, even where topographical knowledge remained imperfect. This process is now being extended out into the seas, while air and space travel has introduced pressures for a third frontier above the earth's surface.

The Romans were keenly aware of the political-geographical significance of the frontier. In the Republic's expansive days the Roman core was protected by a wide frontier zone of client territories under its hegemony, extending as far as Roman diplomatic and military influence could stretch. As romanization spread and the impetus of expansion slackened, the later Empire surrounded itself by a defence-in-depth frontier zone with clearly defined linear limits. This made use of natural features, such as the Rhine and the Danube, but also man-made defensive walls, like the *limes* across southwest Germania or Hadrian's Wall in North Britain. In the steppe and desert of North Africa and Asia Minor, however, lines of defended positions were built to deter incursions of mounted insurgents. Nevertheless Rome sought to maintain a glacis of client tribes forward of these lines as a sponge. Such a policy became less tenable as its enemies learned Roman techniques. Lattimore (1937), in his study of the Great Wall of China, saw walls of that type, including the Roman equivalents, as a means of marking limits for expansion in an attempt to prevent an over-reaching of one's resources and capabilities. As defence they have been notoriously ineffective, perhaps one reason for questioning that as their primary purpose. Lattimore saw the position of the frontier as a compromise, in a sense, between the depth of penetration into the margin of possible conquest, and the ability to hold and administer effectively such conquest.

Political territories emerging from the disintegrating Roman Empire were minor kingdoms around small, scattered groups of settlements separated by extensive tracts of waste, virtually unpeopled except for outlaws, transitional belts from one group to another, its inaccessibility and inhospitality providing something of a natural defensive filter. The broad marcher no man's lands initially offered considerable scope for colonization and expansion of the emerging feudal lordships without undue risk of a clash with neighbours. Hofmann (1969) has suggested that in early times Germanic peoples had no concept of the frontier in a linear sense (and no word for such), so that words like *mark* (march) and *forst* (forest) served to describe the divisions between their political territories.

The small earldoms and kingdoms of Saxon England, for example, each had had a nucleus surrounded by a broad marchland of poor country. The deterrent value of these marches allowed East Anglia and Kent, one protected by wet land on its western boundary, the other shielded by dense Wealden forest, to preserve their identity against more powerful neighbours. On the other hand, the more accessible terrain of the English Midlands allowed the rise of the large Mercian kingdom, able to unite different communities. Even at this early stage, linear demarcation was used, as in Wat's and Offa's dykes, marking the limits between Saxons and Welsh.

The complexity of the legal and moral bonds of feudalism on the Continent hampered the development of clearly defined frontiers.

Pounds (1951) noted that the division of Charlemagne's Frankish empire between his three sons at Verdun in AD 843 was done by allocating territories rather than by drawing a frontier in the true sense. Each *pagus* (a territorial unit probably inherited from the Romans) was given to one or other son in a rather arbitrary fashion, though the division was to lay the foundation for the later emergence of France and Germany. Waste separated one *pagus* from another but, as population filled the countryside, disputes over ownership of timber or hunting rights in it multiplied. In the absence of maps, definition of ownership was by reference to natural features, rivers, streams, or hills, as good approximations. Feudal kingship was a matter of justice and lordship, not of sovereignty, so disputes were usually over feudal titles rather than sovereignty, making boundaries difficult to define. In such a situation the nominal frontier between the French kingdom and the Holy Roman Empire had little real meaning and rulers were seldom concerned with where it lay. On rare occasions when precision was needed, local inhabitants were usually asked to declare whose subjects they were.

The march (meaning 'edge' or 'margin') became institutionalized as the middle ages unfolded, providing a special frontier regime. Marcher lords enjoyed powers normally held only by the king as a reward for their responsibilities. Marcher lordships were established against the Welsh and the unruly north of England as a means of consolidating Norman power (Fig. 4.1). In the North German Plain the marcher lords gradually pushed the frontier of German settlement and political influence eastwards, with new marches established as the advance proceeded, leaving an imprint on the regional names of Trans-Elbian Germany. We may also see the Ukraine (meaning 'at the border') in the same light, a vast march to protect the sedentary Slavs against intruders from the steppe. In the eighteenth century the Habsburg established a march, the *Militärgrenze*, along their border with the Ottoman Turks, where settlers received special privileges in return for perpetual military readiness to deter Turkish raids (Fig. 4.2). We may see the early American West as a marchland, spreading against the Indians as settlers filled territory and new states were organized.

Once people and land became significant taxable assets, the old vagueness had to be replaced by exact definition of ownership, a change confirmed in the Treaty of Westphalia (1648) where sovereignty was recognized as the cardinal fact in political organization. When group identity, proto-nationalism, began to displace personal loyalty and feudal servitude, the frontier as a *line* of

Figure 4.1 Marcher Lordships of England

Adapted from Freeman-Grenville, G.S.P., *Atlas of British History*, London 1979, and others.

division took on a new significance. As Pounds (1951) noted, the French assertion that the Rhine was their eastern limit (of merely academic interest in feudal days) by the sixteenth century provoked German counter-arguments. Whereas the German objection in the middle ages had been rooted in a fear of a French king as Holy Roman Emperor, by the sixteenth century a note of nationalism had crept into the debate, with German Rhenish humanists arguing that the essential criterion in defining this critical frontier was language and demonstrating that the divide between French and German speech lay well west of the river. The issue reappeared during the French Revolution, when, primarily for military reasons, the new regime claimed the Rhine as the 'natural frontier' on which French destiny lay.

Figure 4.2 The Habsburg *Militärgrenze*

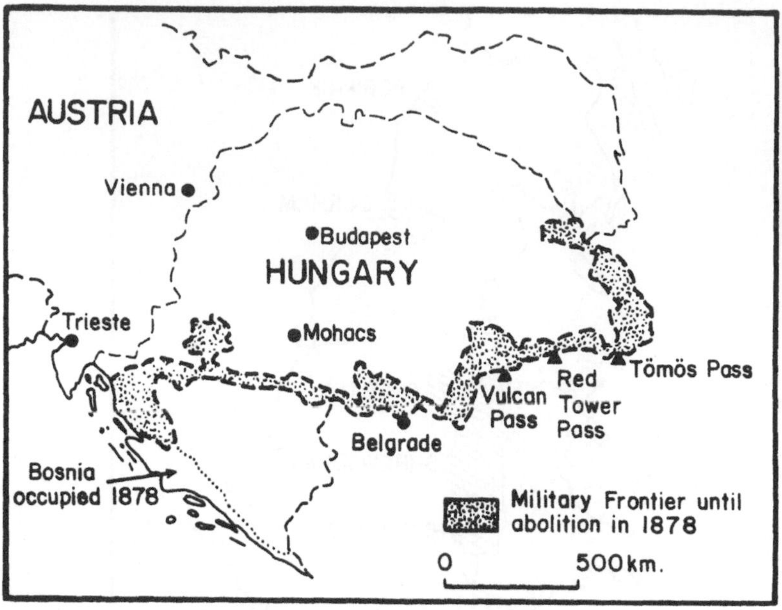

Adapted from *Grosser Historischer Weltatlas*, vol. III, Munich 1981

The French Revolution gave impetus to the nation as a focus of loyalty, intensifying the desire to define territorial limits more exactly. As language and culture became increasingly significant bonds, there was a pressure to know where their territorial limits lay, expressed in the frontier where delineation of boundary lines was made easier by better maps. Nineteenth-century nationalism intensified the preoccupation with careful delimitation and by the twentieth century many frontiers had become virtually sealed limits. Movement across them began to be regulated more and more by passports after the First World War, but possibly the rising mobility and the greater numbers of people on the move made governments anxious to know who was crossing their borders and why. The most divisive and closely guarded frontiers appeared after the Second World War along the Iron Curtain in the 1950s.

DEFINING THE FRONTIER

Terms like 'natural frontier' or 'scientific frontier' seem to imply that some formula can be used to define an ideal frontier. In essence, every frontier (wherein lies a linear boundary) is unique, for nowhere else will exactly the same combination of geographical conditions occur. The line of the frontier is really a compromise between the aspirations of the contending parties, so that successful demarcation of the boundary will depend on finding how grievances on both sides may be minimized, if durability is to be attained. The stability of any frontier may, however, be jeopardized at any time by a change in the power equation between the two parties.

Eighteenth-century philosophers developed the idea that each nation had its 'natural frontiers' beyond which it ought not to extend, but sadly these were not explained by actual geographical examples, though popular opinion usually recognized vague, generalized limits beyond which there was little enthusiasm to go. Certainly the idea is an acceptable one (even if limits are often only vaguely conceptualized) that to extend too far is to overreach one's resources, overtax one's administrative capabilities. Such extension may also incorporate unwilling minorities as citizens. 'Natural frontier' suggests there are features in the landscape which can be readily identified, when really it means 'rational' or 'optimal', or, in many instances, the best line for defence.

Features of the topography are nevertheless used commonly in defining and demarcating the frontier: they appear to be fairly easily described and to be readily identifiable in the field, though in reality this is not always the case. Whereas the frontier in its broadest zonal form may be successfully defined in general physical terms (e.g. 'the northern frontier of Italy lies in the Alps'), the actual demarcation of the boundary by minor features can prove difficult. Watercourses are especially common as boundary markers (possibly because they are readily defined and visible, as well as providing something of a barrier), their use perhaps encouraged by the emphasis given them in early maps, making them appear more important than in reality. The *raison d'être* of modern frontiers of long standing can often be appreciated only by reconstructing the form and perception of the terrain at the time of their inception. The Netherlands-German frontier across the Bourtanger Moor was laid across a waste marchland, though this is hard to appreciate on the Dutch side where the old moor has been almost completely transformed by reclamation, but it is more easily seen on the German side, where more of the moor remains, as pressure to reclaim has been less. Boundaries are often for short stretches laid along man-made features with a linear character such as roads, canals, or railways. The usage goes a long way back, with remains of Roman roads used in Dark Age times (an example is the use of Roman Watling Street during the Danelaw).

National self-determination this century shifted the emphasis to frontiers respecting ethnic and national identity. Such divisions are seldom clear cut and commonly transgress topographical features which might otherwise be seen as admirable markers for boundary lines. The ethnic quilt has sometimes been so complex, however, as to defy delimitation of a clear boundary as in the Banat and Bačka between Yugoslavia, Rumania, and Hungary after 1919. Use of the historical boundary of Bohemia as a new national frontier in 1919 ignored the ethnic divide, incorporating into the Czechoslovak state a large and vociferous 'Sudeten' German minority. After the Second World War, draconic measures to simplify the ethnic map in Europe were marked by mass expulsions, most notably of long-standing German populations from Poland and Czechoslovakia, to produce more clear-cut national divides, while a similar action occurred at Partition between India and Pakistan.

It has been unusual for an international boundary to be closely aligned to a divide in economic geography, even though major territorial claims to whole provinces have frequently been made in economic terms. In many instances boundaries have been fixed well before the distribution and magnitude of mineral or other resources have been appreciated. The Franco-German boundary of 1871 in Lorraine took little interest in the distribution of the minette iron ore deposits since at the time these had little economic value.

In fixing boundaries, considerations of defence have commonly overridden all other criteria. After the First World War, Czechoslovakia accepted a large Magyar minority in southern Slovakia in order to control a key strategic railway and a similar situation arose in western Rumania. The Belgians' interest in Eupen and Malmédy in 1919 was governed by a desire to push the Germans back from overlooking them from the top of the Eifel 'glacis'.

Territorial claims and consequent frontier adjustment can lie moribund for long periods, awakened by a shift in the political balance of power. Afghanistan, for example, pressed for a revision of its border of 1877–88 in 1947 when it knew Pakistan was embarrassed by internal problems and poor relations with India. Danish pressure mounted to reopen the question of its frontier in *Sudslesvig* in 1945 in the confusion of the German collapse, notably unfortunate as the boundary had been fixed by a fair and well-managed plebiscite in 1920. In the heated emotions at the end of wars, victorious powers have short-sightedly commonly fixed boundaries in their own favour, failing to remove the cause of grievances and even generating new ones — the frontiers fixed by the Treaty of Versailles after the 1914–18 War created as many new difficulties as they solved old ones.

THE NATURE OF BOUNDARY DISPUTES

Prescott (1965, 1975) identified four types of frontier or boundary dispute. At the largest scale, claims to whole provinces involve major frontier adjustment. Such a case is the long-held Chinese grievance over the Russian occupation of the Pre-maritime Territory in the Amur and Ussuri basins, seen by China as a product of 'unequal' treaties made in the 1850s. There are also disputes at local level over the actual positioning of the boundary, arising from such things as differing interpretations of treaty conditions or incorrect survey and demarcation of the boundary line. Prescott quotes as an example the French acceptance of the boundary delimitation between their Niger territory and British Nigeria early this century until they found the only practicable route for their military patrols between Niamey and Zinder ran partly through British territory denied to them. Positional disputes may range from quite local disagreements at one or two places to demands for revision of the whole length of a boundary line.

Disputes arise over apportioning natural resources in the frontier zone, with water supplies and flood control notably common as a source of disagreement. An example was the dispute between Britain and the United States in 1900 over tapping Lake Michigan to supply water to carry away Chicago's diluted sewage to the Illinois river. The amount drawn off lowered the level of the lakes associated with Lake Michigan, so reducing the draught and consequently the capacity of lake steamers. Britain was successful in its claim against Chicago for reducing the navigability of boundary waters and the volume of water withdrawn was steadily reduced until 1938. Lowering of the groundwater on the German bank to the detriment of farmers through the construction on the French bank of the Rhine of the Grand Canal d'Alsace caused considerable friction in the 1930s and since the 1950s the German Democratic Republic has been in dispute with the Federal Republic over its effluent polluting the latter's reaches of the Werra. A happier situation arose in the Saar Agreement of 1957 that allowed France to mine coal under German territory from pits on its own territory for twenty-five years. Frontier disputes may arise where changes disrupt old patterns of life, though a relatively less common form. House (1959) pointed to the effect of the change in the

Franco-Italian frontier in 1947 which left communities on either side of the new boundary with ill-balanced pasturage. A special convention was produced to allow an exchange of grazing rights across the line.

Claims for frontier adjustments are frequently argued legally on historical rights or the interpretation of treaties, but sometimes they are made simply on grounds of the appropriateness of the change. For example, Rumania claimed the whole of the Banat on the argument of its natural unity, despite the historic province's ethnic complexity, with strong Serbian representation. Czechoslovakia demanded the urban district of Teschen (Těšín–Cieszyn) because its railway network included the best mainline route across the Carpathians into Slovakia. The Polish counterclaim suggested the Czechs use the alternative railway along the Vlara valley, but this was dismissed because of the heavy gradients and sharp curves.

PLEBISCITES

Frontier changes have usually been made with little or no reference to the wishes of the local population: Lord Russell, speaking in 1864 noted 'the Great Powers had not the habit of consulting populations when questions affecting the Balance of Power had to be settled'. For example, the Germans took Alsace and Lorraine from France in 1871 without any local consultation, particularly in debatable communes along the language boundary, and the French acted likewise in 1919 when they reoccupied the provinces. In 1938, Britain and France acquiesced in Germany's redrawing the boundaries of Czechoslovakia on its claim that the Sudeten Germans wished to be part of the *Reich*, although no inquiry was made of local German or Czech sentiments in the affected territories. There have been, however, instances where better counsel has prevailed and local wishes have been registered, if not acted upon. Such *plebiscites* have usually been restricted to small areas of little strategic importance, where the dominant powers have felt able to permit such a liberal exercise. For areas accorded a plebiscite there have been many equally deserving but denied it. A plebiscite requires careful management if it is to be impartial and a true reflection of local feeling.

Two contrasting plebiscites were held after 1919 to decide frontier disputes between Germany and its neighbours. That held in Schleswig in 1920 was in a largely rural area with some modest towns, of which Flensburg was the largest. In districts of mixed settlement

German and Danish national feeling was fairly clearly differentiated, but there were no major contrasts in living standards. The final result showed polling districts voting with usually clear majorities and the pattern was a good lead to demarcating the new boundary line (Fig. 4.3). Denmark would like to have gained Flensburg and Germany to have retained Tondern (Tønder), but the new division was readily accepted and did not drastically disrupt local life. Agitation, mostly from the Danish side (though with some German opportunist elements — so-called 'Bacon Germans'), sought to reopen the issue after 1945 but quickly died away.

Figure 4.3 The Schleswig plebiscite of 1920.

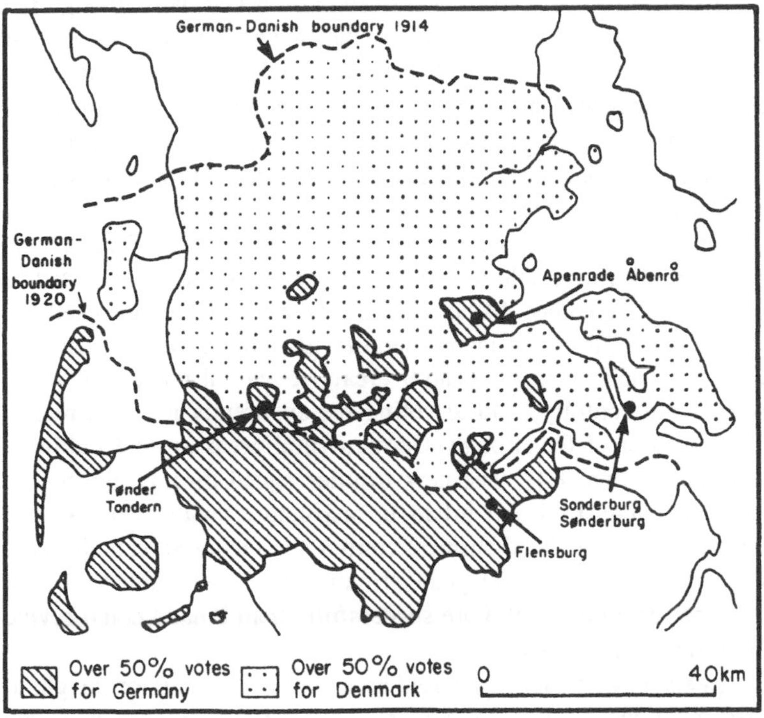

Adapted from *Topographischer Atlas von Schleswig-Holstein und Hamburg*, Neumünster 1979

The plebiscite area in Upper Silesia comprised both farming country and industrialized districts with substantial towns on the coalfield. Over the west and north the countryside was largely German-speaking, but in the centre and southeast the rural people spoke mostly Polish.

The industrial towns were overwhelmingly German in character, containing both local Germans and immigrants from other parts of the *Reich*, but there were also Poles, both local and immigrant in origin. Many townspeople, originally Slav in speech, had accepted German language and culture, but there was considerable bilingualism and many families spanned German and Polish cultural sentiments. Despite the confusing ethnic mix, the population was predominantly Roman Catholic. Whether German or Polish in sentiment, most people felt their future would be economically more secure in Germany.

While those of German sentiment voted in the plebiscite of 1921 overwhelmingly to remain in Germany, at least 40 per cent of those of Polish sentiment voted likewise. In the plebiscite area as a whole, the majority (60 per cent) voted to remain in Germany, though the pattern varied considerably from one part to another, with 56 per cent voting for Poland in the centre and as much as 71 per cent in the extreme southeast. In the industrial towns at least 55 per cent voted to remain in Germany. A simple division of territory along nationality lines was impossible because of the extremely confused areal pattern of voting. The final division was a compromise between a British proposal (favourable to Germany) and a French one (exceedingly favourable to Poland). Even so, the area allocated to Poland still contained 44 per cent of its voters who had opted for Germany, while a not inconsiderable pro-Polish vote was left in Germany, notably in the central area. At several places territorial awards ran contrary to local voting for apparently no good reason (e.g. transfer of the strongly German rural areas around Lublinitz to Poland) (Fig. 4.4). The new frontier sliced through the highly integrated industrial and urban infrastructure, with everyday life seriously dislocated and business confidence shattered on both sides. The richer and better managed German economy was able to sustain the area remaining to it more successfully than Poland could revive the territory it received.

An elaborate treaty supervised by a special commission sought to ease relations across the new boundary and help maintain the infrastructure intact, but the situation deteriorated as chauvinism intensified in Germany after 1933 and following Pilsudski's death in 1935 in Poland. Although voting had been supervised by an Inter-Allied Commission and was regarded as secret and well-conducted in a 90 per cent turnout, the division left much ill-feeling and accusations of vote-rigging, intimidation, and other malpractices in both countries.

Small transient territories under special management may exist

Figure 4.4 The division of Upper Silesia, 1920–1922.

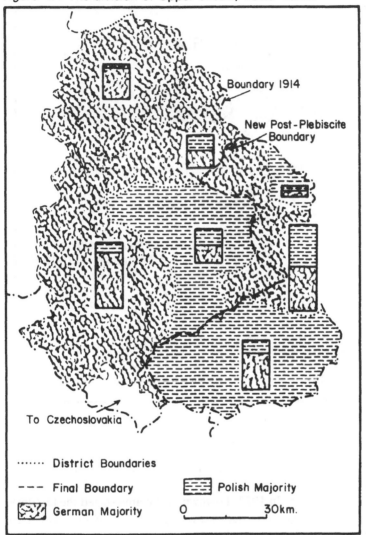

The columns are proportionate to the numbers of voters in the plebiscite and are divided according to the proportion of votes. Various sources

while decisions over frontiers are made. In 1919 French forces held the Memelland until its future status and the boundaries of Lithuania could be agreed. The 1914 frontiers of Germany had excluded Lithuania from a long coastal strip of territory of mixed German-Lithuanian character, but the intention of the Versailles Treaty was to institute some sort of autonomous territory (possibly like the League

Figure 4.5 Trieste after the Second World War

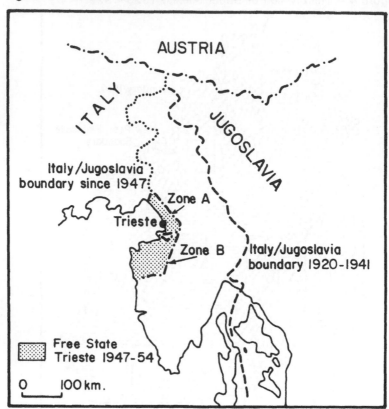

Compiled from various sources

of Nations' arrangement at Danzig). Although there was no formal consultation, there appeared to be little popular desire for incorporation into Lithuania. In 1923, however, Lithuania seized the territory without French opposition and its sovereignty was formally recognized in 1924, although it accepted under pressure a Memel Statute conceding a measure of autonomy and right of appeal to the League over disagreements. In 1945 alignment of the Italo-Yugoslav frontier closer to the ethnic divide met difficulties over Trieste (Fig. 4.5). The 1947 Treaty of Paris set up a neutral, demilitarized Free State around the city under United Nations supervision. Disagreements between the Anglo-American administration in the city and so-called Zone A and the Yugoslavs in Zone B resulted in the scheme collapsing. After lengthy negotiations the Free State was divided between the Italians in the city and most of Zone A and the Yugoslavs in Zone B plus some of Zone A, while the 1954 London Agreement marked an Italian guarantee to maintain a free port and protection for minorities on both sides.

ARBITRARY AND PROVISIONAL BOUNDARIES

International boundaries may be created arbitrarily by precipitate action or unanticipated events. Such 'accidental' boundaries, commonly applied with little consideration for their impact on the human landscape, are consequently lines of tension and inherent instability, invariably virtually 'closed' frontiers. The Polish seizure of Vilna and 'Central Lithuania' in 1920 imposed a quite arbitrary frontier that remained completely closed as Lithuania refused any contact with Poland whatsoever. The boundary between the two German sovereign states was conceived in 1945 purely as a line of demarcation between Allied occupation forces, part of a four-power administration for the whole of Germany, and fixed almost entirely along pre-existing local government boundaries. During the Cold War it developed into a closely guarded, almost 'impermeable' frontier, cutting ruthlessly through villages, farm holdings, and even individual houses, dislocating communities and local relationships. Equally arbitrary is the so-called 'Green Line' in Cyprus, established subsequent to the Turkish invasion of 1974, which has caused exceptional dislocation not only along its length but also to the whole economy and society of the island.

'Provisional' frontiers may similarly become fixed without careful delimitation, as between Poland and Germany on the Oder-Neisse rivers. The boundary was established in 1945 temporarily until a proper line could be agreed and demarcated in a 'final German peace treaty', though over forty years later that treaty is as far away as ever. It has become accepted in default by both sovereign German States through agreements designed to ease relations with Poland — by the German Democratic Republic in the Görlitz Agreement of 1950 and by the Federal Republic in the 1970 Treaty of Warsaw. Even good intentions may not succeed — Lord Curzon in 1918 proposed a well-conceived line for a Polish-Russian frontier, but the Soviet-Polish war ended in a far less geographically satisfactory frontier in the 1920 Treaty of Riga.

GENETIC DEVELOPMENT OF BOUNDARIES

Definition and demarcation of political boundaries may take place at different stages in the development of the cultural landscape (Fig. 4.6) At what stage and under what circumstances the boundary line is fixed is often significant when examining political geographical problems. The boundary line may well have been fixed when the territory was largely uninhabited: such an *antecedent* boundary (Hartshorne 1936) is in anticipation of problems when advancing frontiers of settlement meet. Such pioneer boundaries are those between Canada and Russia under the Anglo-Russian treaties of the 1820s and the 'Forty-ninth parallel' between Canada and the United States, established and modified under treaties between 1782 and 1846. At the time it was fixed, topographical knowledge was incomplete and the territory inhabited by only a few settlers and trappers, apart from the Indians, for whom little concern was shown. Antecedent boundaries, established by usage before the major features of the cultural landscape emerged, have a largely pragmatic and historical character, as in so many instances where they arose from early marchlands in Europe.

Boundaries may be defined *subsequent* to development of the cultural landscape, like those between France and Belgium or Holland and Germany. Such boundaries frequently coincide in general with major or minor divisions of natural or cultural regions, for which the term *consequent* has been proposed. In recent times the emphasis has been on defining boundaries to approximate closely to ethnic patterns. Sadly, treaties following wars have reordered boundaries with little if any reference to the cultural landscape, deservedly designated *superimposed*. Whereas the *subsequent* boundary can be described as concordant (i.e. running wiht the grain of the cultural landscape), the *superimposed* boundary commonly disregards such order and may be said to be discordant. The boundary between the two sovereign German states since the 1950s is a classic example of this type, but others were the boundaries imposed on Poland in the eighteenth-century, partitions that lasted until 1914 in modified form. The plebiscite is one mechanism for making a new boundary fit the cultural landscape but, since the Second World War, in superimposing boundaries, an artificial concordance has been attempted by population transfer to simplify the ethnic pattern and create, if possible, a sharp ethnic divide along the new frontier, well seen along the 'provisional' frontier of Poland on the Oder–Neisse rivers.

When the line of an international boundary is moved, the differences in the cultural landscape along the original line remain

Figure 4.6 The genesis of frontiers and boundaries

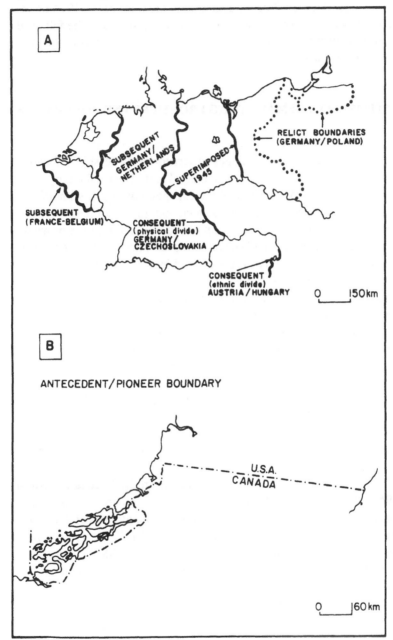

visible for a long period, marking the course of what may be termed a *relict* boundary. In Poland, even today, the lines of the 1914 boundaries between Russia, Prussia, and Austria can still easily be followed in architecture and many other facets of the built environment and economic infrastructure.

DEFINING AND DEMARCATING FRONTIERS AND BOUNDARIES

Complex diplomatic negotiations conducted by statesmen and diplomats usually prefix changes in territorial ownership, of which the frontier and boundary form an integral part. The preliminary stage is generally a broad definition of the changes envisaged, with *allocation* of provinces, towns, or known natural resources between contending parties, followed by a more careful *delimitation* of where the new boundary should in general terms lie. These written statements are translated into reality on the ground by a boundary commission which defines the new border in the field by exact references and undertakes a *demarcation*. In well-mapped and developed country the commission generally starts work immediately broad agreement has been reached, so that an exact document describing the course of the new boundary and any conditions for its management can allow early institution of legal sovereignty by the countries involved. In pioneer or colonial territory, however, a practical need to establish the new limits exactly may not arise for a considerable time. Commissions working in developed territory usually have a precise brief to follow, but in pioneer country they are allowed considerable discretion to deviate within defined limits in the field to allow ease of demarcation or administration.

A commentary on the process comes from the Vienna Award in 1940, when parts of Rumania were transferred to Hungary. The diplomats had reputedly drawn and erased so many lines on their map that the draughtsmen producing the fair copy for the treaty document had to make calculated guesses at several places. Hitler's interpreter, Schmidt (1951), makes a similar point over the new boundaries for Slovakia in 1940, when 'pencil lines on the draft map were several kilometres wide in reality'.

PROBLEMS OF DEFINITION AND DEMARCATION

The terms of definition must be unambiguous and non-contradictory, not always achieved. In developed country, use of place and feature names must be accurate and precise, especially where alternatives exist, with names taken from maps checked in the field. In 1944 allocation of German territory to Poland was confused because two Neisse rivers exist. The Soviet negotiators referred to the western (Görlitzer) Neisse, whereas the Anglo-American officials thought of the proposals in terms of the eastern (Silesian) Neisse, but nobody at the time seemed to question whether both were speaking of the same river. As the Russians got their way, the Polish frontier advanced westwards by almost two hundred kilometres.

Watercourses are commonly used as boundary lines, being readily visible and usually easily described in treaties. Confusions can, however, arise: in defining the US-Canadian boundary in Maine, British and American claims identified different sources of the St Claire river until an agreed source was fixed in 1798. There is often the question of whether the actual line should be a median line along the *Talweg*, should lie along the main navigable channel or even on one or other bank. Furthermore, how should changes in the river's course be catered for? Along the Rio Grande on the US-Mexico boundary the shifting course has required constant surveillance and on several occasions renegotiation, especially since sudden floods alter the river's meandering, braided bed, making definition difficult. The practice had been for the boundary to follow 'the normal channel' and to move with the gradual shift of the meanders through normal erosion. A sudden change (as when a flood cut through a meander neck) left the boundary unchanged, following the abandoned bed. In 1905, as this practice had led to complications, a treaty amendment agreed that, where such changes in the course left territory of either state on the opposite bank, such parcels should pass into the sovereignty of the state on that bank. Exceptions were made for areas over 250 ha in extent or inhabited by more than two hundred people. In recent times regulation and straightening of the course have seen adjustments to the boundary to conform with the middle of the new course. The problem was solved differently on the Upper Rhine under a treaty of 1827 between France and Baden. The boundary was agreed to be 'the most favourable route for downstream navigation during the ordinary period of lowest water'. If dispute arose over two arms of the river, the one with the deeper soundings would be used as the boundary, but each year the line would be resurveyed and its course marked by posts in October following the period of high water. The line so marked would remain unaltered until the next survey, irrespective of any changes of course meantime. On occasion where the boundary has run along streams or small rivers, the line has been commonly fixed on one or other bank, as cited by Adami (1923) along the Biala-Przemsa and Pilica at the Third Partition of Poland in 1796, defining the boundary between Austria and Prussia. Islands in rivers present problems, especially as the median line of the *Talweg* used as the boundary may shift from one side to the other through natural change. In the Ussuri dispute between Russia and China the islands in question are claimed by China as being on their side of the median line of the main navigation channel under the 'unequal treaty' of 1860, whereas the Soviet claim to the islands is based on a Russo-Chinese Agreement of 1861 that laid the boundary along the Chinese bank of the river. After the First World War most treaties accepted the median line as the boundary on non-navigable rivers, while the middle of the principal channel was the line on navigable rivers. This is less simple than it appears, since the distinction between navigable and non-navigable rivers is not always straightforward.

It is customary to apportion large lakes by dividing them with a median line, though exactly how it is drawn is open to several techniques (Boggs 1940). Some large lakes have, however, been divided by straightline boundaries along meridians and parallels. The first technique was used on the North American Great Lakes, the second on major African lakes. Treaty arrangements usually assure free navigation and fishing rights, as do those agreed by the countries bordering the Bodensee in the nineteenth century. More recently treaties have covered the use of waters and control of pollution. Where one country holds parts of both banks, the waters as encompassed are under direct and exclusive sovereignty (e.g. Lake Khanka in Siberia). Small water bodies are commonly put completely within the territory of one border state.

Treaty makers have favoured relief features for defining frontiers but, though they appear clearly on maps, serious difficulties of inter-pretation can emerge on the ground. In ill-surveyed territory, relief and drainage features often turn out on careful survey to be radically different from travellers' reconnaissance reports. Mountains are commonly favoured, because they have qualities for defence and appear to be barriers, while they are often a divide between national groups. Yet they have usually considerable width, commonly lack an easily defined crestline and may have one that is not always coin-cident with the watershed. The difficulties can be exacerbated by com-plex patterns of intermontane valleys with equally complex human patterns, all making problems for field surveyors.

The treaty of 1881 fixing the boundary between Chile and

Argentina allowed differing interpretations when applied in the field. It stated that 'the frontier line shall be marked . . . by the highest summits of the said chains (Cordillera) which form the watershed'. The Argentinians interpreted this as meaning the actual boundary should follow exactly the watershed (which was well removed in places from the line of the highest peaks). The Chilean view maintained it was not the watershed but the main peaks which were the key to interpretation and the watershed should be applied only in relation to those peaks. In 1902, with British arbitration, what was essentially a compromise was reached. A similar dilemma has also underlain the Sino-Indian dispute over the so-called McMahon Line in Assam.

Inadequate knowledge of the topography created considerable difficulties in defining the Canadian-Alaskan boundary. Under the 1825 Anglo-Russian treaty Russia was to hold a coastal strip south to 54° 40' N as far inland as the summits of the coastal range. The negotiators had, however, recognized that their knowledge was incomplete by inserting a clause to the effect that, if the coastal range summits were more than 30 miles (48 km) 'from the ocean', the boundary line should be parallel to the windings of the coast at a distance inland never more than 30 miles. When discovery of gold in the Yukon aroused Canadian interest in crossing the coastal range, by that time in American hands, difficulties were at once apparent. Improved topographical knowledge showed the coastal ranges to be cut into sections by fjords penetrating as far as 140 miles (225 km) inland from the outermost coastal islands. The Canadian view was that the boundary followed this *interrupted* range, so putting the heads of the longest fjords in their hands for port construction. The Americans argued that the treaty intended the westernmost *continuous* range behind the fjord heads. A joint commission between 1903 and 1906 agreed a boundary running from peak to peak but nowhere further inland than 30 miles.

Definition of the boundary line in mountain country along a so-called 'military crest' was adopted by the commission delimiting the Bulgarian-Turkish frontier of 1878. This followed neither the summit nor the watershed, but was marked along pronounced breaks of slope of value as defensive positions. As a consequence, the more level surfaces of the passes and saddles remained in the hands of one state.

Hill country generally offers few clear summit lines or readily identifiable watersheds, so boundary definition is commonly made along watercourses, between individual named features or along elements of the man-made landscape. Because of their defensive value, defiles, ridges, or modest scarps become particularly important. In this manner the escarpments of Lorraine were used in delimiting the frontier after the Franco-Prussian War of 1870. Minor relief features play a significant role in open plains, so the Dutch-German border shows a general association with the high terraces and *Stauchmoränen* overlooking the Roer valley, while the modern Belgium-Netherlands border reflects the old march in riverine marshes, and the Russo-Polish frontier has long been influenced by the vast Pripyat marshes.

Treaty-makers may take the apparently simple expedient of straightline boundaries from given coordinates, saving the costs of detailed survey and delimitation. The technique was used as early as 1494 when the Pope made a division 370 leagues west of Cape Verde between the Portuguese and Spanish empires under the Treaty of Tordesillas. This type of boundary is mostly used in little known country, but it has sometimes been imposed in developed territory in crisis situations, as in the Thirty-Eighth Parallel to divide Korea in 1945 or the Seventeenth Parallel division between north and south in Vietnam in 1954. Straightline boundaries have been common divides between colonial territories in Africa and in the desert interior of the Arabian Peninsula, but they are also met in the Americas: the most famous is the Forty-Ninth Parallel between the United States and Canada. Experience from the United States–Canada joint boundary commission suggests a straightline boundary is probably as costly to administer as a normal sinuous border adapted to the natural landscape. A rather special case was the arc-shaped boundary between Germany and France after 1871, drawn around the major French fortress of Belfort to keep it out of German artillery range.

THE FRONTIER AS A DIVIDE

The frontier line is inevitably a divide and barrier, but how far it acts in this manner will be coloured by relations between the countries on either side. Though the actual boundary line is always indicated by intervisible markers (posts, cairns, or the like), one or other or both states may erect barriers (barbed wire, ditches, walls, minefields, etc.) to impede contact and restrict movement over the line to a few official crossing points. The ultimate 'ugly frontier' is the Inter-German border, where on the Democratic Republic's side is an electrified and booby-trapped fence, barbed wire, ploughed and mined strips, all ground cover cleared, watched from manned towers (Fig. 4.7). Behind all this lies a 500 metre-wide prohibited access zone and a further 5 km-wide belt where access and residence are carefully supervised. In contrast, between members of the European Economic

Figure 4.7 Guard zones on the Eastern side of the Inter-German Frontier.

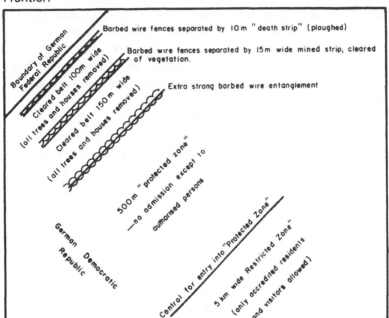

This is perhaps the world's most intensively guarded frontier

Community or between good neighbours like the United States and Canada, identification and surveillance of the boundary is in an exceedingly low key. In some parts it is difficult to identify the actual line. Along the Franco-German border the relaxed conditions of to-day contrast with the 1930s when there was rigorous control on both sides, as well as formidable defence works and long lines of 'dragon's teeth' tank traps.

Perceived as an insecure periphery, the frontier zone has usually suffered from a low investment confidence, with serious implications for economic development and social infrastructure. The border with the Socialist *bloc* has been so disruptive for the German Federal Republic that along it stretches a 40 km-wide special assistance area, the *Zonenrandgebiet*. In contrast, the easing of international mobility within the European Economic Community has made frontier locations attractive for member countries to penetrate each other's markets (e.g. Dutch Limburg or *Saarlorlux*) or for outsiders to establish branch plants just across the border within the European Economic Community (e.g. Swiss branch plants just across the French and

German frontiers near Basel). A controlled border can have serious consequences on long-standing patterns of nomadism and trans-humance. Modifications were made to the new Polish-Czechoslovak frontier in 1919 in order not to disrupt seasonal movement of flocks in the Carpathians. This has also been widely experienced in Central Asia, notably between the Soviet Union and China, while the implications on the Swiss-French frontier in the Jura were interestingly explored by Daveau (1959).

The frontier influence on movement has an effect on transport networks. Even across open frontiers, movement is encouraged only at official crossing places: where relations are strained, these may be few indeed. Not all crossing places handle all types of traffic, so influencing overall patterns of movement. Usually there are special arrangements for frontier residents within a given distance on either side, important for trans-border labour movement, as across the Franco-Belgian border. Austria's new boundaries in 1919 cut across several railway links between provincial towns and Vienna, with special non-stop 'privilege' trains allowed to use the old routes across what had become Hungarian or Italian territory. After 1919 the railway between Eupen and Malmédy ran partly through Germany as a narrow corridor of Belgian territory and was given special extra-territorial status. Transport routes have in many instances been built over difficult terrain rather than follow an easier trajectory involving crossing an international boundary. In West Africa the French Guinean railway from Conakry to Kankan was laid over the awkward Futa Djallon mountains rather than follow an easier and more direct course across British Sierra Leone.

MARITIME BOUNDARIES

Most states have some coastline and consequently an interest in sovereignty over adjacent waters for naval defence and for unimpeded access to the high seas. Roman legal commentators conceived the idea of sovereignty over the sea in their argument that the emperor had the right to punish wrongdoers at sea in the manner he punished them on land. In the middle ages it was asserted that the Holy Roman Emperor had the right to resist communal use of the sea by appropriating parts of it, by granting a privilege, or through long and unchallenged use of certain waters. A fourteenth-century Italian lawyer argued that rulers owned any island within a hundred miles of their shores and had authority over the intervening waters, an assertion

Figure 4.8 Maritime boundaries I.

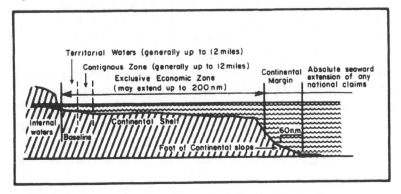

Exploitation of marine resources has encouraged claims to sovereignty over the continental shelf

Figure 4.9 Maritime boundaries II.

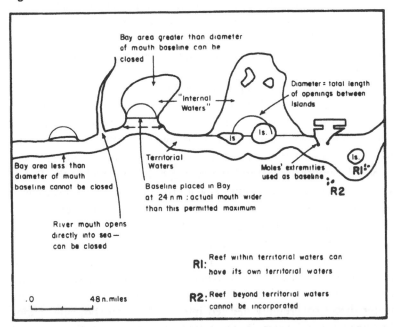

Coastal indentations and other features create numerous difficulties in defining a state's territorial waters

used by Venice to claim waters between the islands it held and the heads of deep bays in the northern Adriatic. In northern waters, even by the thirteenth century, there was already a tradition of control by different groups of parts of the North Sea and even Atlantic waters. The Norwegians, Danes, Scots, and English, by proclamation, treaty, or long-established custom, exercised control over imprecisely defined parts of these seas. By the late fourteenth century greater commercial and fishing activity put strains upon this usage but, despite mounting friction, only in the late sixteenth century did legal arguments emerge on the rights of countries to adjacent waters. Elizabethan English lawyers maintained the Queen had jurisdictional rights to the middle of seas between her kingdom and foreign countries, while another claim was to the horizon seen from the shore (variable depending upon the height of the shore). With control over Greenland in 1585, the Danish Crown (united with that of Norway since 1380) claimed control over the whole ocean and the Skagerak. The Danes freely issued licences to other nationals to sail in these waters and to fish, but they did not apply to a coastal belt of waters around Danish-controlled coasts, defined as two leagues wide in 1598 (eight nautical miles by Danish reckoning) and later extended for four leagues. Subsequent changes, mostly reductions in width, were made under pressure from others. Spain and Portugal notably sought to prevent free use of the high seas, claiming absolute sovereignty over large areas of ocean during their imperial ascendancy in the sixteenth century.

Grotius, a sixteenth-century Dutch lawyer, in his *Mare Liberum*, argued that no state could control the open sea, but this was contested by his English counterpart, Selden, in his *Mare Clausum*. He claimed that seas around the English coast were 'owned' by the English Crown, but the subsequent growth of English overseas seaborne commerce and imperial ventures seems to have swung opinion round to the Dutch view. The seventeenth century saw a growing conceptual awareness of territorial waters and especially of their importance in event of war. They emerged as extending up to three nautical miles (one league) offshore, the so-called 'cannon-shot rule', a generally accepted measure. Though rather vague and variable as a measure, by 1730 it had become embedded in international law. By the late eighteenth century the Scandinavian idea of territorial wates as a continuous zone under direct sovereignty merged into the 'cannon-shot rule' under Mediterranean influence. By the mid-nineteenth century, three nautical miles as the extent of territorial waters had been generally accepted.

Most important was the concept of 'innocent passage' under which any vessel might sail freely in any territorial waters so long as it had no hostile or detrimental intent. States with large merchant and fishing fleets and considerable navies found it in their interest to keep territorial waters to the minimum, to allow 'innocent passage', and to make the greatest possible areas of open sea free for navigation, yet the three-mile limit was a minimum and there was no agreed maximum. Spain in 1760 claimed six miles and the Russians, worried at foreign incursions for seal and walrus hunting, set their territorial waters between 10 and 100 miles, according to the waters. The first international conference at The Hague in 1930 failed to reach any agreement on the extent of territorial waters and the question remains open. As new independent countries have emerged, the problem has grown more complex, so that some Andean American countries claim 200 miles and once conservative states have pushed their territorial waters to 12 miles or more.

The baseline from which measurements are made forms the maximum seaward margin of a nation's internal waters (bays, inlets, estuaries, etc.) as well as the inner limit of territorial waters and recently defined zones of surface waters and seabed beyond. The baseline should be measured from low water as indicated on hydrographic charts and should follow the sinuosity of the coast with provisions made in respect of bays, estuaries, and other indentations, besides coasts of extreme irregularity. Pronounced indentations in coastlines must have a 'closing line' across their natural seaward entrance to simulate a 'normal baseline', helping to smooth the line of coast and making position fixing easier for ships at sea. Coastal indentations may be 'mere curvatures of the coast' and may be considered for closure only if the line of closure serving as the diameter of a semicircle makes such a semicircle smaller than the water area enclosed. Regardless of configuration, no line of closure across the mouth of any indentation may exceed 24 nautical miles. If the mouth is wider than 24 miles, a baseline of that length must be drawn within the bay to enclose the greatest possible water area. Where a bay has islands across its mouth, the accumulated water distance of the baselines between those islands may not exceed 24 nautical miles. Wide estuaries also count as bays, though small river mouths may have a baseline drawn across them where they empty into the sea, while permanent harbour works may be regarded as part of the coast, thus extending the baseline seawards.

Bays and other waters may be claimed on historical grounds and not defined by the rules above, but formal claims for recognition must demonstrate that sovereignty has been exercised for a considerable period during which it has been recognized and accepted by other countries. Historic bays include the Spencer Gulf, Hudson Bay, Bristol Channel, the Chuckchi, Kara and Laptev Seas, and Chesapeake Bay.

Islands are entitled to their own baseline and territorial waters, generally according to the same rules as mainland coasts. Their baselines may overlap those of adjacent islands or the mainland and by this means archipelagic states generate territorial seas of their own. Land above water at low tide but covered at high tide, if it lies within the territorial seas of a state, may extend those waters seawards, but such low tide elevations may not, however, have territorial seas of their own if they lie beyond such waters.

Along particularly sinuous or jagged coasts it is permissible to join headlands by straight baselines. The arbitrary baseline established is used to measure territorial waters, though it may not deviate to any great extent from the general direction of the coast, but this vague directive has been liberally interpreted. The seas lying to landward of such a baseline must be closely associated with the mainland domain as internal waters. The individual legs of the baseline should not exceed 24 miles, though this has been frequently ignored. In 1951 the International Court of Justice accepted a Norwegian baseline, reputedly drawn in 1869 and 1889, comprising 47 segments along the coast; 47 of its 48 base points rested on offshore islands or low tide elevations, with only the remaining one on the mainland. The segments themselves varied in length from a few hundred yards to over 40 miles and the baseline at one point lay 15 miles from the nearest land and at another was 18 miles from the mainland. Some countries invoking the straight baseline principle in recent times have made excessive claims: despite its relatively smooth coast, Albania has claimed such a baseline; and both Indonesia and the Philippines have used such baselines to encompass vast areas of the high seas into the internal waters of their archipelagoes.

A newer concept of the 'contiguous zone' has effectively extended territorial waters by giving a country jurisdiction over defined functions such as customs and immigration. Some have, however, argued for exclusive fishing rights in the zone, allowing concessions to friendly countries. The concept began as a further nine nautical miles beyond the three-mile territorial waters, but as so many countries declared these to extend to 12 nautical miles, the contiguous zone has become an entitlement to a further 12 nautical miles. Where the coast of two states face each other across a narrow sea, the territorial waters and any contiguous zone will extend to the median line drawn from the baselines to equidistant points in the sea.

Traditionally and still acceptable in international law, the 'high seas' beyond territorial waters have been free and open to all. Nevertheless, states have sought to encroach on open waters and reluctant recognition has conceded some extensions of sovereignty. The precedent was the Truman Doctrine of 1945 that claimed for the United States all the natural resources of the seabed and subsoil of the continental shelf contiguous to its territory. No sovereignty has been acknowledged for the waters above the shelf. This was essentially confirmed by the 1958 Law of the Sea Conference when the legal edge of the continental shelf was fixed at the 200 m isobath (countries equipped to do so might exploit beyond that depth) and later the sub-sea limit was redefined at 200 nautical miles from the coastal baseline where the continental margin does not extend so far, or otherwise to the outer edge of the continental margin. The latter is often difficult to determine precisely, so considerably different interpretations have been made of it.

The so-called 'exclusive economic zone' allows sovereign rights over essentially economic functions such as exploration of the sea and seabed resources across a sector of ocean up to 200 nautical miles from the coastal baseline. It also permits the erection of artificial islands and the right to control pollution. Freedom of navigation has, however, to be guaranteed for all vessels on the high seas within the zone. This development cast a new importance on small oceanic islands, for these could acquire sovereignty through the 200-mile rule over a zone covering immense sea areas exceeding 125,000 nautical square miles. Any land which is 'naturally formed and above water at high water and can sustain human habitation or economic life of its own' is accorded the right to a continental shelf (if it exists) and to an exclusive economic zone of its own. Isolated rocks unable to sustain habitation are entitled only to a territorial sea and contiguous zone, sufficient to make small, remote infertile islands or new volcanic islands significant. An important proviso, agreed with the exclusive economic zone concept, gives landlocked countries transit rights across territories between them and the sea. They are also entitled to participate in exploitation of the *living* resources of the exclusive economic zone of coastal states adjoining them on an 'equable basis, taking account of the relevant economic and geographical circumstances of all the states concerned'.

Where international boundaries run to the coast the countries involved need to agree a line of division between their territorial waters (and contiguous zone). Treaties and agreements in many instances have failed to explain adequately how the seaward prolongation of the boundary should be defined. The usual methods have delimited territorial waters between countries by a line whose direction has been fixed *either* in relation to the land boundary *or* in relation to the coastline. The first case makes the seaward boundary a projection of the land boundary from its last straightline section. The second case considers the seaward boundary independently except at its point of contact on the coast with the seaward projection perpendicular to the general direction of the coast, but on a ragged coast this is open to several interpretations. This latter method was used in the settlement of the Norwegian-Swedish dispute over the Oslo Fjord by the International Court of Justice in 1909 and was said to be in keeping with a treaty of 1661. Because both methods can be to the disadvantage of one or other party, Boggs (1940) proposed construction of a median line from points on the coast, a complex technique useful where there are offshore islands.

THE THIRD FRONTIER

After Blériot's successful crossing of the Channel in 1909 the British expressed concern over the control of national air space. The subsequent rapid development of aviation brought international discussion at a conference in Paris in 1919 and others in 1928 and 1929, but most importantly at Chicago in 1944, Montreal in 1947, and The Hague in 1955. International law of the air embodies much closer control over freedom of flight than there is over navigation at sea, for in the air there is no 'innocent passage'. The so-called 'Five Freedoms' agreed in 1944 are severely constrained by the right of sovereignty over national air space. Use of air space and establishment of civil air services rests on bilateral agreements between states which can prescribe corridors for flight, define prohibited zones, and other conditions. Air routes for commercial aviation are consequently not always the shortest great-circle trajectory, resulting in expensive detours.

Space-vehicle development raises the question of how far national air space extends upwards. So far powers capable of putting vehicles into space have done so without regard to national air space below their orbits. Some jurists have argued for a vertical limitation to national air space, rather on the analogy of national sovereignty over the sea. On the analogy of the 'continental margin', it has been suggested that air space should be defined in relation to marked levels of physical change in the atmosphere (little easier, if at all, than defining the continental margin). Possible levels might be the stratopause (40–50 km altitude), around 80–90 km where gravity begins to be felt, or at the lowest altitude for satellite orbits (110–160 km). The United Nations treaty governing activities in space of 1967 gave guidelines for the exploration and use of outer space, including the moon and other celestial bodies.

5

Geographical aspects of national defence

A prime responsibility of the state is to ensure the security and welfare of the nation in its territory, and part of this remit lies in providing military defence against hostile powers. National defence has strong geographical elements, since time and space are the essentials of strategy and tactics. It also requires careful assessment of national economic resources and manpower, especially in the formulation of a grand strategy to attain specific political objectives. War is constantly changing in its nature, so defence plans must be continually revised, with wide-ranging territorial, social and economic implications.

Though countries may seek to solve international disputes by diplomatic means, negotiations and posturing in themselves have been regarded as unlikely to succeed without a veiled threat of armed intervention in event of deadlock or failure. Armed conflict has been seen as the ultimate resort if all other persuasion fails, embodied in Clausewitz' view that war is the continuation of policy by other means. Armed conflict at any level should not be viewed as a self-contained phenomenon but rather as a part of a broader spectrum of political grand strategy. Though few people would accept the historian Treitschke's view of war as an elixir to revitalize nations, many countries have acted on the soldier's dictum that attack is the best defence, with war frequently precipitated by a pre-emptive strike to forestall a possible enemy move. In war the key to success is in maximizing the qualities of time and space through surprise and innovation. There is, of course, no set scenario that leads to war — the fortuitousness of events may trigger it at any of several different stages of escalation. The point at which diplomacy ends and war begins has depended historically on the perceptions, values, and philosphies of the contending parties in the social and political milieu of their age.

NATIONAL DEFENCE

The strong spatial dimension of national defence has considerable impact on the organization of national territory, though the details of each defence system will depend on the size, shape, and terrain of the territory concerned, on its relationships between land and sea, as well as on its human and other resources (Fig. 5.1). Because territory is one of the most sacrosanct of all national possessions, defence is usualy organized so that none may need to be surrendered, even temporarily, to an attacker. This is particularly important for small states whose chances of survival depend often upon holding off an invader until help from powerful friends can arrive. Exceptionally, however, Russia has made use of its vast territory in a defence strategy of luring an invader ever deeper into the interior, stretching and weakening his lines of supply until he may be easily repulsed and thrown back. Claims to territory, though made ostensibly on historical or ethnic grounds or like reasons, have usually had a strong strategic dimension, so that the term 'natural frontiers' has generally been a euphemism for militarily advantageous frontiers. In laying out defence, particular protection is given to key industrial or food-producing districts and especially to the capital city. Capture of an enemy capital is usually regarded as a claim to victory (consider the competition between Soviet and Anglo-American armies for the capture of Berlin in 1945). Protection of economically vital districts is of particular concern where they are vulnerable to enemy attack near the frontier, as demonstrated in France between 1870 and 1940. Plans may include designation and preparation of some highly defendable area as a 'national redoubt' in the event of a military disaster (as was planned in Yugoslavia in the 1930s).

A notable defence component is usually built into economic and population policies. The spatial expression is seen in such things as location of armaments industries in 'safe areas' in interior districts (as the Poles tried to do in the Kielce-Radom triangle in the 1930s); in land use and settlement (planting of woodland and location of settlement were controlled by the military in the frontier areas of France from the seventeenth century); in the modification of transport systems by military needs (the Roman road network or railway networks in nineteenth-century Europe). Strategic considerations formed an element in the design and function of the large economic planning regions, charged with creating a high level of local self-sufficiency, introduced in the Soviet Union during the late 1930s. Between the wars manpower planning for military needs principally underlay

Figure 5.1 Model of the defence of national territory.

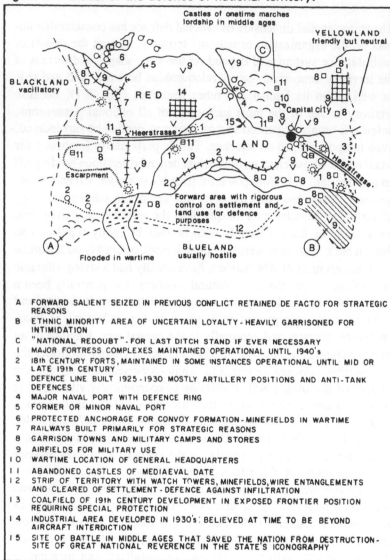

A   FORWARD SALIENT SEIZED IN PREVIOUS CONFLICT RETAINED DE FACTO FOR STRATEGIC
    REASONS
B   ETHNIC MINORITY AREA OF UNCERTAIN LOYALTY - HEAVILY GARRISONED FOR
    INTIMIDATION
C   "NATIONAL REDOUBT" - FOR LAST DITCH STAND IF EVER NECESSARY
I   MAJOR FORTRESS COMPLEXES MAINTAINED OPERATIONAL UNTIL 1940's
2   18th CENTURY FORTS, MAINTAINED IN SOME INSTANCES OPERATIONAL UNTIL MID OR
    LATE 19th CENTURY
3   DEFENCE LINE BUILT 1925-1930 MOSTLY ARTILLERY POSITIONS AND ANTI-TANK
    DEFENCES
4   MAJOR NAVAL PORT WITH DEFENCE RING
5   FORMER OR MINOR NAVAL PORT
6   PROTECTED ANCHORAGE FOR CONVOY FORMATION - MINEFIELDS IN WARTIME
7   RAILWAYS BUILT PRIMARILY FOR STRATEGIC REASONS
8   GARRISON TOWNS AND MILITARY CAMPS AND STORES
9   AIRFIELDS FOR MILITARY USE
I O WARTIME LOCATION OF GENERAL HEADQUARTERS
I I ABANDONED CASTLES OF MEDIAEVAL DATE
I 2 STRIP OF TERRITORY WITH WATCH TOWERS, MINEFIELDS, WIRE ENTANGLEMENTS
    AND CLEARED OF SETTLEMENT - DEFENCE AGAINST INFILTRATION
I 3 COALFIELD OF 19th CENTURY DEVELOPMENT IN EXPOSED FRONTIER POSITION
    REQUIRING SPECIAL PROTECTION
I 4 INDUSTRIAL AREA DEVELOPED IN 1930's: BELIEVED AT TIME TO BE BEYOND
    AIRCRAFT INTERDICTION
I 5 SITE OF BATTLE IN MIDDLE AGES THAT SAVED THE NATION FROM DESTRUCTION -
    SITE OF GREAT NATIONAL REVERENCE IN THE STATE'S ICONOGRAPHY

The diagram seeks to identify the main spatial elements in national defence
since Napoleonic times

population policy in several European countries, while difficulties
caused by labour shortages in the Soviet economy, arising from
maintenance of excessive military forces into the late 1950s, subsequently forced a reduction in their establishment in favour of the
civilian economy.

Airpower has radically recast war, changing long-established values of territory and terrain, but also adding a new dimension to seapower. Military and naval success, whether in attack or defence, now largely depends on command of the air. The old advantage of deep interiors beyond easy interdiction has been substantially eroded, with defence of very large territories made more difficult through the range and mobility of the air weapon (including most recently missiles), giving rise to an idea of the 'embarrassment of space'. Naval defence and seapower remain crucial for many states, especially those heavily involved in overseas trade. Seapower was, of course, particularly significant in the development of the great European empires overseas in the eighteenth and nineteenth centuries, but it nowadays remains a key component in the global strategy of the superpowers.

Over the last forty years, a complicating factor in all defence systems at whatever scale (national, regional, continental, or global) has been the apparently insoluble imponderables posed by nuclear weapons. The dilemma faced is how to give at least some protection in event of crisis escalation across the nuclear threshold.

As the conduct and technology of war have changed over time, so the territorial and resource aspects of national defence have had to be adapted. The changing needs of defence have a considerable impress upon the spatial organization of national territory, leaving clear marks on the landscape, both from existing and relict installations. Though the nature of war has been in constant change, seven distinct phases may be distinguished.

The Roman age

The Roman Empire held together through its military superiority and co-ordinated system of territorial defence which responded to the changing political situation and to the balance of power between it and its enemies. Defence of the vast territories had to be achieved with modest manpower resources, but the secret of Roman success lay in superior organization. The Romans understood the importance of strategic and tactical mobility, provided by a good command structure, the excellent road network providing good communications by an elaborate system of couriers and signal stations. Most importantly, aware of the subtleties of deterrence and its limitations, they appreciated that the dominant dimension of power was more psychological than physical. When their enemies came to realize this last fact, the fate of the Imperium was sealed.

In the expansive phase of the Empire, the political structure was notably hegemonic and depended on a defence in depth along fluid and ill-defined frontiers. Client states, overawed by Roman power and kept loyal by threat or subsidy, acted as a sponge to absorb attack, fearing that failure to perform this task would bring the wrath of the legions. The legions were generally stationed well back from the frontier in case they were needed for action against internal troubles, dominating major routes and effectively dividing the Empire into manageable units. The calculated risk accepted was that not all the frontier would be under attack at one time, so reliance was on mobility and the ability to concentrate legions wherever and whenever serious trouble appeared.

The heavy, slow legions, able to exert a relentless pressure, were capable of ultimate siege or assault tactics. Their visible presence was most important in the west against simple tribal subject peoples, whereas the advanced eastern client states better appreciated Roman potential, so there was less need to keep legions physically present. The legions were most intimidating where sedentary populations had fixed assets, but on the desert frontier in Africa and the Near East they were less effective, not only tactically but also because of logistical problems. On the desert frontier the Romans developed cavalry, modelled on local forms, but never distinguished in performance. The great problem was to keep an adequate central strategic reserve and legions had to be moved as occasion demanded, though the spatial pattern of the Empire made strategic redeployment cumbersome and slow.

By the first century AD, to economize on manpower, especially along the lengthy Danube frontier, fixed border defences became increasingly significant in an otherwise mobile strategy (Fig. 5.2). The old 'surge' tactics were replaced by a sustained defence capability, as the frontiers became virtually static in the second century AD, manned often by 'native' troops. Creation of these defences and their associated infrastructure was a massive investment spread over three hundred years. The defence works were primarily deterrent to low-intensity attacks and major threats were still countered by concentrated mobile forces carrying the counter-attack on to the 'glacis' beyond the defences. The frontier walls and palisades offered some protective security for civilian life and helped to isolate frontier populations from dissident influences beyond. In North Africa and the Near East continuous walls were less easy to construct and maintain and less effective against the enemy and were replaced by defended oases.

The Roman Empire (like all others) faced the dilemma of how

Figure 5.2 Types of Roman defences.

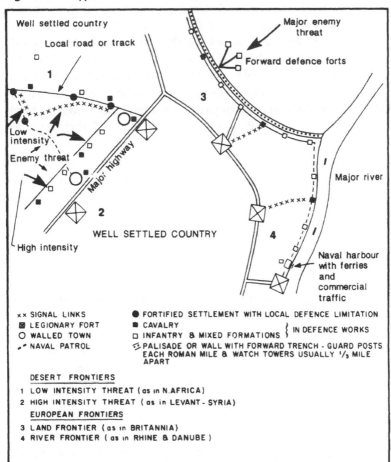

Imperial defence systems were adapted to local conditions. Adapted from Luttwak, E.N. *Grand Strategy of the Roman Empire.* London 1979

far expansion could be pushed before resources, especially manpower, were overstretched. At any point on the frontier this assessment rested not only on local conditons but also on events elsewhere. Even by the first century AD effort was being made to stabilize the limits along lines easy to dominate, as reflected in the elimination of the awkward salient between the Rhine and Danube of the Agri Decumates, sealed off by a wall. Large conquests by Trajan in the Near East were quickly abandoned as Hadrian tried to strengthen the western borders. As Rome's enemies became more proficient and persistent, they penetrated deeper in their raids, so fortifications were built further

109

into the interior. The emphasis shifted from an offensive capability to a defensive role in tactically dominant positions such as those astride main roads. Towns, villas, even farms, began to be fortified, able to hold out until help arrived, not unduly difficult as the enemies had little or no siege capability. Provincial security was increasingly sacrificed to the security of the Empire as a whole, with considerable political repercussion, especially as by the fourth century AD militia and peasant soldiers were given growing responsibility. From the third century, the threat of seaborne raiders who often left their boats to raid far inland created a new burden for hard-pressed defenders. The Empire was now seriously overtaxed, with resources in men and wealth insufficient to face an ever more efficient enemy, so retreat began from peripheral provinces in a vain hope of saving the imperial core.

The middle ages

War was perhaps more the normal condition of feudal society than peace, with shifting loyalties a prime cause of conflict. Kings fought each other but also had to maintain their authority by fighting their own lords, who also squabbled with each other, while piracy and brigandage added to the difficulties of maintaining law and order. Kings relied on the honouring of feudal obligations to recruit their forces for, as simply first among their peers, they depended on the latter's loyalty for their power. Though kings were as rich as, if not richer than, their most powerful vassals, they seldom commanded wealth enough to support more than a personal force of retainers, so when war threatened they had to muster their underlings. The lords and knights brought their own arms and an agreed number of fighting men, forming a reasonably equipped and practised force. The peasantry could be enlisted as *levée-en-masse* for a notional period (usually forty days) and thereafter the king's charisma extracted longer service. An army of such ill-armed, ill-trained reluctant warriors was a motley rabble difficult to manage and the real fighting force was the mounted knights and their retainers (often bowmen). By the thirteenth century, however, commutation of feudal service into payments allowed kings to recruit specialists, with a growing body of men from several classes ready to devote themselves to the profession of arms. Infantry came back into its own, well seen in the English longbowmen in the Hundred Years' War, where they proved devastating against the French, still recruited on the older feudal

110

tradition and much dependent on heavy mounted knights.

War in the middle ages was dominated by fortified places, especially castles, also considered essential because of a lack of power-ful central authority to enforce law and order. The secret of a king's power was to hold the key castles, either through his own possession or through his most trusted vassals. The difficulties of moving armies over meagre roads intensified when these routes were overshadowed by menacing castles from which dangerous sallies could be made and no real command of the countryside could exist without their con-trol. As a consequence, war was mostly long and usually inconclusive sieges, attempts to force castles into submission by starvation or disease rather than by frontal attacks. It was not until the Crusades that reasonably efficient siege machines were developed, though these were countered by increasingly sophisticated fortifications until the coming of gunpowder in the fourteenth century marked their slow decline. The most important castles were usually those guarding the frontier marchlands of the kingdoms, not only for defence but also as bases for expansion. The powerful castles of the Welsh and Scottish marches of the English kingdom are a good example, with those of 'Little England beyond Wales' around Pembroke as the great Anglo-Norman springboard for the invasion and subjugation of Ireland. Expansion into a marchland often involved colonizing settlement in the form of fortified towns or 'bastides' seen in North Wales but also in southern France. Away from the marchlands castles often had little strategical purpose and were more displays of lordly wealth, like many built in England without licence during the virtual anarchy in the reign of Stephen (1135–54).

War at sea rested on merchant ships carrying fighting men. Vessels in the Mediterranean with lateen sails and oars were capable of more mobile naval engagements than the northern cog that simply sought to get alongside an opponent for a hand-to-hand fight. Introduction of artillery at sea brought broadside engagements, where luck counted more than good gunnery. Alfred the Great in the ninth century had nevertheless realized the importance of naval defence in his struggle against the Vikings and other English kings had followed, with the Cinque Ports under obligation to provide ships for the naval defence of the south coast. Even from late Roman times the English coast had had formal defences, for there was a constant fear of invasion from the sea. Naval engagements were often quite formidable, like Sluys in 1340 when 147 English and 190 French ships clashed, and even remarkable amphibious operations were mounted, as witnessed by the Norman invasion of England in 1066 (fortunately facing little

111

opposition at sea) or the successful landing of the Earl of Derby in Flanders in 1337, opening the Hundred Years' War.

Early modern period

The shift to a monetary economy meant kings could 'buy' soldiers, and by the fourteenth century the concept of the 'king's army' emerged in Angevin England, so that war became seen as the 'sport of kings'. As a sign of prestige kings were willing (though not always able) to pay for ever larger armies, but their generals were less enthusiastic, for larger armies were harder to control, especially as logistical skills did not keep pace — nothing is less successful and more difficult to manage than a hungry army. In 1430 the English fielded 6,000 men at Agincourt; by the sixteenth century, armies numbered as many as 30,000 or 40,000 men a century later. The eighteenth century saw armies of 70,000 or 100,000, while Louis XIV (1643–1715) drained the French economy with an army of 400,000. Marshal de Saxe in 1757 emphatically considered 46,000 men as the ideal number to manage, for 'multitudes serve only to perplex and embarrass'.

The rank and file came from the poor (said to 'thieve, turn soldier or starve') and the officers from the gentry and aristocracy. Such armies were held together by draconic discipline and rigorous drill — Frederick the Great said that if his soldiers were allowed to think, none would remain in the ranks. Desertion was high, often as much as 60 per cent during campaigns, with high losses to illness and fatal wounds through a lack of hygiene (especially in the Thirty Years' War). Kings often preferred to employ professional mercenaries, most famous of whom were the Swiss pikemen and crossbowmen. The French favoured mercenaries, claiming one was worth three men — one more soldier for France, one less for the enemy, and one Frenchman to remain at home to pay taxes. Nevertheless, once unemployed or when pay failed, mercenary armies often became a severe nuisance by turning to brigandage.

With armies having to be paid, infantry came into its own, for cavalry was expensive and at a disadvantage as missile weapons were increasingly used, besides the difficulty on occasions in finding enough horses. The armies were moved as one long, slow column and this cumbersome form took much time to draw up in battle order, while the quality of the troops made formal drills and formations essential. Mercenaries also favoured formalized drills, since as professionals they had a strong sense of self-preservation: rules were formulated to reduce the mortal danger of combat by making clear when one might concede victory with honour. Manoeuvre rather than head-on clash was pursued and battle declined unless conditions were exceedingly favourable, with engagements plentiful but decisive battles few.

Crude logistics made feeding armies difficulty, while foraging gave too good an opportunity for desertion, so interest shifted to prestocked magazines. Late in the eighteenth century, the magazines at Berlin and Breslau held sufficient grain for 60,000 men for two years, some of the stores being over forty years old. Tied to these supplies, an army dared not venture more than 160 km away from its magazine. There was sometimes an attraction in sending one's army into somebody else's territory to relieve home food supplies, although in the early eighteenth century only five areas in Western Europe were capable of provisioning an army without magazines — the Rhineland and Westphalia, parts of France, the Spanish Netherlands, and Lombardy. Limited mobility was exacerbated by poor roads and the long civilian tail of camp followers ministering to the army's needs. Because of the difficulty of moving in winter, operations were in a low key from October to April, so the relatively short campaign season made it important to bring a reluctant enemy to battle. A decisive victory had to be won early in the season if there was to be a chance of following it through.

The problems of movement, manoeuvre, and supply, as well as a desire to keep 'vicious' fighting to a minimum, made generals favour lengthy sieges. Improvement in artillery, however, forced the development of new types of fortifications, seen in the seventeenth century in new low-profile defences able to resist cannon shot, notably refined by the French engineer Vauban. The low, sloping, and usually earthen ramparts resisted cannon balls and were too steep to storm, while rings of these ramparts supported each other, and corner bastions could enfilade attackers. The French surrounded their vulnerable northern and eastern flanks by two lines of such defences, begun about 1678, while many more were built by German princes, notably in the Rhineland (Fig. 5.3). Because storming these forts was generally unprofitable, it became an established usage that after about forty-eight days of siege, if the besiegers had not been dislodged, an honourable capitulation could be made. The importance of these works was reflected in several peace settlements that included the garrisoning of forts beyond one's own boundaries or the demilitarization of forts in a broad zone facing one's own frontier, such as forts held in the German Rhineland by the French or the Dutch garrisons in a line of 'barrier forts' across southern Belgium as a defence against

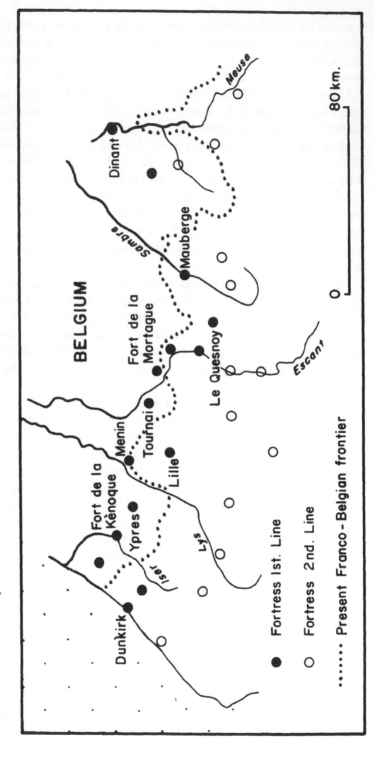

Figure 5.3 Vauban's defence of France. The vulnerable northern border was protected by two lines of forts planned in the late seventeenth century

France under the Treaty of Ryswick of 1697.

From the sixteenth century defence at sea saw the emergence of proper naval forces distinct from armed merchant ships, though piracy remained common and privateering was quietly encouraged. As European powers became increasingly involved in overseas colonial ventures, the importance of seapower grew. Spain and Portugal were quickly challenged by France, England, and the Dutch, with varying fortunes in the competition for naval supremacy. As the Iberians declined, for a brief time Holland performed well, but the silting of the Dutch estuaries, which the technology of the age could not counter, undermined their position, preventing the use of deeper-draught, more seaworthy vessels able to carry more armament as the English and French were doing. France's commitments as a continental power hampered devotion of the necessary resources to the navy and by the mid-eighteenth century supremacy had effectively passed to the English, though France remained a serious adversary.

The age of Napoleon

The French Revolution changed the whole tenor of national defence as the armed forces became identified with the people and not the king. The popular citizen army of the French, even if initially inexperienced, had a new *élan*. Emergence of national identity in increasingly coherent form also emphasized the desire to consolidate territorial holdings and define legal sovereignty behind linear boundaries which could be resolutely defended. The Revolution's *levée-en-masse* produced an exceptionally large French army of over 600,000 men, eventually to rise to over one million under arms by 1800, a level of mobilization never previously experienced and widely felt among the population. By organization and tactical innovation, Napoleon sought to maximize the value of this large force, ending the rather limited warfare characteristic since the early seventeenth century, and greatly improving his mobility by abandoning magazines and making the army live off the country, possible because desertion was not such a problem in this citizen army. Reducing the army's tail by eliminating the last camp followers, Napoleon made artillery (once handled by civilian contractors) solely the responsibility of army personnel, achieving economies that represented 5,000 additional cavalrymen or several hundred extra guns in an army of 100,000 men. Movement was eased by a major road-building programme, extending into captured lands and copied by some German princes.

Mobility and tactical manoeuvre were improved by dividing the army into self-contained corps, each a miniature army, with mixed divisions of cavalry, infantry, artillery, and supporting arms, able to fight its own limited actions until help arrived. By appreciating the importance of good communications and signalling, Napoleon could disperse his forces, the corps moving in widely spaced columns, easing pressure on roads, food, and forage, bewildering the enemy as to where the main thrust would come until the last minute. The earlier pattern of sieges and manoeuvres to avoid decisive engagements shifted to seeking out the enemy to engage him in battle on Napoleon's own terms.

The Revolutionary war drew military and civilian interests together as not previously experienced. The shift to aggressive tactics brought numerous decisive battles and greatly increased casualities, even if victory was attained. Well before the horrific losses of the 1812 Russian campaign, manpower shortage was beginning to be felt, but more critical was the inability to make good the loss of horses that impaired Napoleon's cavalry and artillery as late as Waterloo. The Napoleonic campaigns reflected national defence dilemmas as war-weariness in France cooled popular support for the war, aggravated when Napoleon's fortunes began to falter as he failed to maintain his innovative tenor in strategy and tactics. Once these became stereotyped, his bluff was more frequently called as his enemies developed unanticipated countermoves. A significant development for the future was Napoleon's recognition of the importance of the economic element in war (his 'Continental System'). Another 'modern' feature was the way guerrilla warfare of the Peninsular campaign (the embarrassing 'Spanish Ulcer') tied down large French garrison and lines-of-communication forces.

Though little advance came in naval warfare, the Napoleonic period confirmed the importance of command of the sea as the British fleets kept the French navy from linking up with its allies. French aspirations in the Middle East came unstuck through the British naval victory at Aboukir Bay and the British navy was an absolute deterrent to Bonaparte's most cherished dream of invading England, a project whose fate was sealed by Nelson's triumph at Trafalgar in 1805.

The age of iron

The new technology of the Industrial Revolution made war possible on an enlarged and more destructive scale, formidably exacerbating the dilemmas of defence. Improvements in firearms greatly increased fire-power, with breech-loading making possible the use of rifles from the lying or kneeling position. Rifling in the barrel gave a more accurate trajectory and higher muzzle velocity increased range and penetrating power. Applied to artillery, these qualities were increased by the use of explosive or armour-piercing shells, while more accurate fire over greater range meant guns could be withdrawn from the immediate field of battle. The greater fire-power and the lessons of fieldcraft learned in colonial wars (especially in North America) saw abandonment of the old formal drills and pitched battles of charge and counter-charge, while the bright uniforms gave way to mute colours. Effort was made to improve fortresses to withstand the new weapons and the most important were developed as clusters of forts spread over an arc of 40–50 km, but even so their role steadily declined, most coming simply to house garrisons.

National defence was revolutionized by the mobility brought by the railway, able to move men and equipment on a grand scale over great distances, formidably accelerating speed of deployment or concentration. The importance of the railway was first demonstrated in the American Civil War, when the Union's superior railway system compared with that of the Confederates proved decisive. The lesson was repeated in the Austro-German war of 1866, and again in the Franco-Prussian War of 1870 when German railways moved 400,000 men compared with 100,000 by the French. Not a little railway construction, particularly in Europe and Asia, had powerful strategic undertones. The steamship, too, made possible quick and reliable movement of troops across the oceans, well illustrated by the Boer War of 1902. Hand-in-hand with this the electric telegraph and the telephone were to provide a major advance in communications between field armies and the general staff.

Naval defence changed radically as steam vessels loosened the old dependence on wind and sea conditions, allowing development of naval ports at sites unsuitable in the days of sail. Steam warships needed, however, more sophisticated facilities and 'coaling stations' took on a formidable strategic importance. The clearing of masts and rigging from decks allowed breech-loading guns to be mounted in revolving turrets, so the old broadside tactics could be modified.

Military affairs intruded more into civilian life as the growing size of armies was sustained by compulsory military service in most countries, providing large pools of reservists readily mobilized. Indoctrination of conscripts became a significant influence on public opinion. To supply the large armies, the armaments industry became a growing sector of the national economy, but whereas earlier the difficulty of moving artillery pieces saw gunsmiths located near likely theatres of war, manufacture of the much more sophisticated weapons of the nineteenth century moved to the main industrial areas, preferably well behind the frontier. Industrial countries equipped themselves with their own national designs of weapons from their own industries, but they also sold freely to friendly countries.

The combination of industrial technology and vast conscript armies was first demonstrated in the Great war of 1914–18. It drew in the total human and material resources of the belligerents, with manpower becoming a key element in national defence as a dilemma emerged as to how the needs of the armed forces and industry could be satisfied. To provide these numbers long term, the need for a 'population policy' introduced a new aspect of government, though the problem had been identified before the war when, in 1911, an estimate showed that the French would need to mobilize 83 per cent of their available man-power to attain the same size of army as the Germans would muster with only 53 per cent of theirs. The whole political-military balance had been tipping in Germany's favour since the 1850s through the slow growth of French population compared with that of their more ebullient neighbour. As the war progressed, a rising proportion of industry was devoted to supplying the vast armies and careful planning of resources undertaken to surmount critical bottlenecks. Economic warfare became a key element as the Germans sought almost successfully to cut off Britain's seaborne supplies by submarine warfare, pursuing the campaign so vigorously that it precipitated American entry into the war. The Allied blockade of Germany, however, bit deeper, creating food shortages that contributed substan-tially to the collapse of 1918, when, though the military situation was far from hopeless, the home front could stand no more.

On the main battlefields of France and Flanders generals seemed unable to master the management of immense fire-power and gigantic armies. All mobility was lost as a slogging match between two fixed lines of trenches developed, where in battles like those of the Somme and Marne horrendous losses were incurred, squandering the belligerents' manpower and sapping their industrial strength. A new dimension was added as aircraft demonstrated their potential over the battlefront and on raids behind the lines (especially the long-distance raids by German airships) where industrial installations became legitimate targets. A second new dimension was the motor vehicle (even though armies depended heavily on horse transport until the end in 1918) whose potential was first appreciated when the French rushed troops to the front in Paris buses and taxis. The armoured fighting vehicle, the tank, was a striking innovation which had it appeared earlier and in sufficient numbers, might well have been able to break the stalemate of trench warfare.

Once it was realized that the Great War had not been 'the war to end wars', national defence again became a major issue, but more difficult because of manpower shortages and economic impoverishment. The search for arms limitation was abortive, though some progress was attained in naval affairs, notably in limiting the number of capital ships. These large and expensive vessels, symbols of national prestige, had not been particularly successful during the war, when naval operations were dominated by destroyers, fast cruisers, and submarines. Though it had not become a major weapon, poison gas raised apprehension, with measures to outlaw it (and possible biological weapons) in the Geneva Protocol of 1925, although research and development continued.

Land defence was the main dilemma, with the choice between a vast fortified line on the model of the trenches of the Western Front or reliance on tanks and aircraft in the hope of bringing back mobility. Desperately short of manpower, economically embarrassed, and uncertain of the resolve of the people for defence, the French government decided on preclusive security in the forts of the Maginot Line, expensive to man and leaving few resources for a strategic reserve, but it could be held by troops of moderate quality (Fig. 5.4). It faced the German border but was not carried westwards from Sedan to the sea, partly because it was believed the Germans would never again attack across Belgium, but partly because it was politically unacceptable to build it along the Belgian frontier, even though Brussels and Paris in no way co-operated in defence.

Many British and particularly German generals saw the other option as most profitable. Limited to a regular army of 100,000 men, the *Reichswehr* (little stronger militarily than Czechoslovakia or Poland), the Germans needed to make this army highly mobile and of good quality, but there were serious constraints imposed by the Versailles Treaty. German officers serving as advisers to the Red Army (a long Russian tradition) experimented in war games deep in the secretive heart of Russia to develop a concept of fluid battle using tanks, motorized troops, and aircraft, *Blitzkrieg*, in which armour supported by aircraft sought a weak point in the enemy front, flooding through when a crack developed.

Some students saw airpower being used primarily against industrial and civilian targets on a massive scale, overshadowing its tactical battle-front use. Defence of interior targets meant providing expensive

Figure 5.4 Interwar defences of the Franco-German frontier.

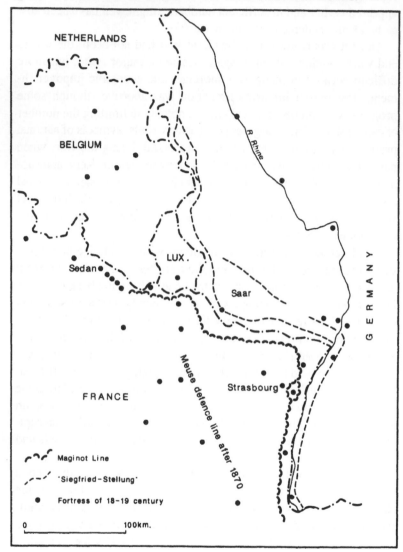

Both France and Germany built elaborate fixed defences, with the most sophisticated works in the French Maginot Line.

anti-aircraft batteries, barrage balloons, fighter protection, and an observer early-warning system. Civil defence would involve mass evacuation from towns, building air raid shelters, and anti-gas measures. How civilian morale would react was an unknown.

Economic warfare now became an element of national defence as much was spent on stockpiling essential raw materials; in seeking substitutes for imports (e.g. synthetic petrol and rubber production in Germany); in dispersing vital industries from likely target areas; and in home food production. Several countries, notably France, Germany, and Italy, instituted population policies clearly aimed to boost growth and replace lost manpower. Governments had in the interest of defence to become much more *dirigiste* in economic and social affairs.

The age of technology

Defence became a major concern for most European countries from the mid-1930s, with expectation of war about 1941 or 1943, but conflict came by 1939. German pressure to redress territorial grievances was matched by Anglo-French 'appeasement', interpreted as a sign of weakness, and exacerbated by a reluctance to rearm as a counter to German and Italian preparations. The new war quickly became even more 'total' than that of 1914–18 and was soon carried into Africa and the Middle East, while the Japanese seized the opportunity of the European powers' embarrassment to spread their imperial ambitions beyond China, where they had been fighting since 1931.

The new fluid, mobile war was demonstrated from the outset by the blitzkrieg tactics of the German victory in Poland, repeated just over six months later in France and Flanders (Fig. 5.5). The Germans outflanked France's preclusive defence, the Maginot Line, by unexpectedly attacking across Belgium again. The French, low in morale, appeared to have no alternative strategy and collapsed, their resolve further undermined by German 'psychological warfare'. The further development of blitzkrieg took place in the vastness of European Russia, where large, heavily defended 'hedgehogs' replaced a conventional linear defence, while the Anglo-American invasion of *Festung Europa* in 1944 was a further refined form.

Airpower came to dominate war as blitzkrieg demanded 'tactical' air superiority, a lesson vigorously applied in the Anglo-American invasion of Europe in 1944, whereas the *Luftwaffe*'s inability to master the skies over the English Channel in 1940 had nullified hasty German preparation for a seaborne invasion of Britain. Airborne troops were an innovation used successfully by the Germans in Holland and Belgium and instrumental in the capture of Crete without traditional command of the sea. Threat of airborne invasion had been taken

Figure 5.5 The *Blitzkrieg* model.

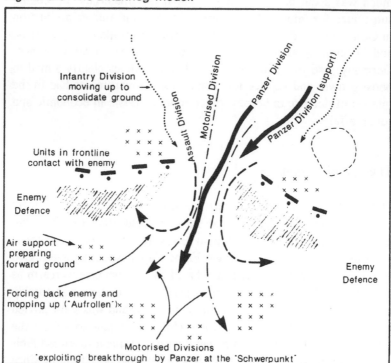

Infantry Division
moving up to →
consolidate ground

Units in frontline
contact with enemy

Enemy
Defence

Air support →
preparing
forward ground

Forcing back enemy and
mopping up ('Aufrollen')

Assault Division

Motorised Division

Panzer Division

Panzer Division (support)

Enemy
Defence

Motorised Divisions
'exploiting' breakthrough by Panzer at the 'Schwerpunkt'

This German concept was the core of tactics in the 1939–1945 war. Adapted from Miksche, F., *Etude sur la Tactique Allemande*, London 1944

seriously in Britain in 1940–2. War was now carried far behind the battlefront into the enemy interior by 'strategic' airpower. The *Luftwaffe* had mounted fierce attacks on British cities in 1940–41, but their magnitude was dwarfed by massive British and American raids on targets in Occupied Europe and Germany. These raids needed vast resources of men and materials and, though highly destructive, failed to halt industrial production or to break civilian morale. There has been much subsequent debate as to whether the effort committed to strategic air forces could have been better employed in support of naval operations or battlefield troops, but there is no doubt they tied up considerable enemy resources in anti-aircraft and civil defence.

Once again, at sea, the German submarines came near to cutting off merchant shipping to and from Britain, even though convoys had been introduced immediately at the outbreak of war. The tide was turned, however, by the use of aircraft, whose ever increasing range and rising attack capability made them valuable submarine hunters.

Major fleet engagements were most characteristic of the Pacific theatre, where aircraft carriers ousted the capital ships in importance, and American naval success in no small part arose from technological superiority over the Japanese.

The Second World War further intensified the interdependence between the civilian and military sectors, and a higher proportion of the civilian population was drawn into the war economy. The increasing involvement of national resources greatly raised the cost of war, while its challenges immensely stimulated technological innovation. The cost was harder to bear than in 1914–18 when the European powers had gone to war at the zenith of their industrial wealth: in 1939 they were still impoverished and their economies had become seriously overtaxed even by 1942. The war ended with the traditional military winners and losers, but political reality was different, for now economics and technology decided the real victors. 'General War', a feature of history since the seventeenth century among European powers, had become so expensive that even the 'great powers' could no longer sustain it from their own resources, and real power shifted to a new class of superpower, of which by far the most powerful was the United States. It was acknowledged that the Soviet Union had also entered the superpower class through the political and military role thrust upon it by the war. The major thrust of the ensuing peace became the effort of the Soviet Union to develop the requisite economic strength to match its new-found political position, while seeking to constrain American military and political predominance by manoeuvre and posturing within the psychological dimension of power.

The power equation between the military and diplomatic was, however, completely changed in the closing phases of the Second World War by two new weapons of unimaginable destructiveness. The first was a true rocket bomb, the German V-2, used devastatingly against London and Southeast England in the very last months of war. The second was the atomic bomb used by the Americans over Japan after the landing on Okinawa against fanatical Japanese resistance had made them anxious to avoid a direct landing on mainland Japan, fearing well over a million casualties or even a repulse. To avoid these risks, it was decided to use the atom bomb to force a Japanese capitulation.

The nuclear age

Within five years of the American atom bombs exploding over

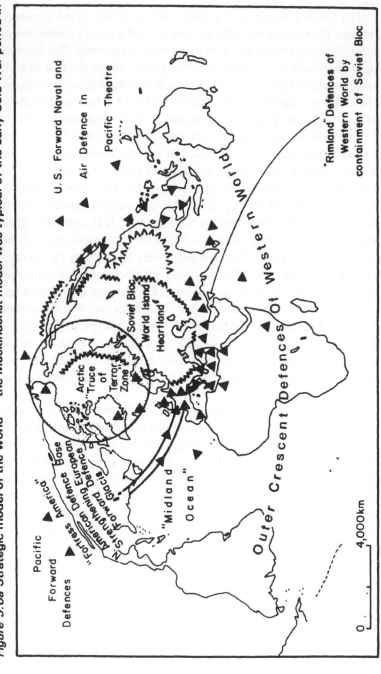

Figure 5.6a Strategic model of the world — the Mackinderist model was typical of the early Cold War period in the 1950s

Pacific Forward Defences

"Fortress America" European Defence

Strategic Glacis

"Midland Ocean"

Arctic "Truce of Terror" Zone

Soviet Bloc 'World Island' Heartland

U.S. Forward Naval and Air Defence in Pacific Theatre

Outer Crescent Defences Of Western World

"Rimland" Defences of Western World by containment of Soviet Bloc

0   4,000km

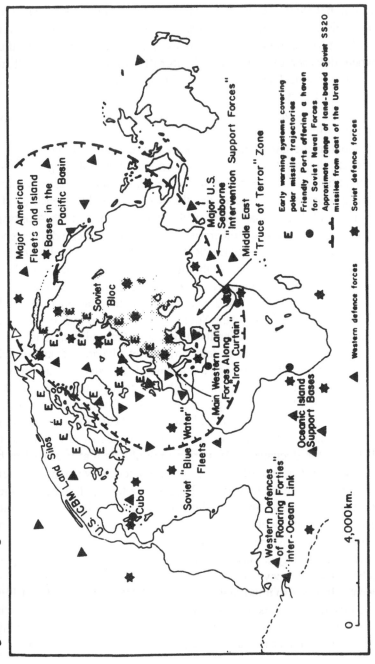

Figure 5.6b Strategic model of the world — a looser strategic pattern using 'hedgehot' defences emerged in the 1960s

Major American
Fleets and Island
Bases in the
Pacific Basin

Major U.S.
Seaborne
"Intervention Support Forces"

Middle East
"Truce of Terror" Zone

Soviet
Bloc

Main Western Land
Forces Along
"Iron Curtain"

Soviet "Blue Water"
Fleets

U.S. ICBM Land Silos

Cuba

Western Defences
of "Roaring Forties"
Inter-Ocean Link

Oceanic Island
Support Bases

E   Early warning systems covering
     polar missile trajectories

●   Friendly Ports offering a haven
     for Soviet Naval Forces

     Approximate range of land-based Soviet SS20
     missiles from east of the Urals

✸   Soviet defence forces

▲   Western defence forces

0   4,000 km.

Japan, a Soviet-made device had been detonated (albeit owing much to Soviet espionage in the West). Britain and France also developed their own devices, with possession counting as much for prestige as defence in the eyes of powers still trailing strands of glory, with de Gaulle believing membership of the 'nuclear club' gave 'a seat at the top table'. The superpowers, however, made every effort to prevent further proliferation of these weapons. The strategic threat intensified when delivery by conventional aircraft was replaced by ballistic missiles (using German rocket technology) carrying several independently targetable warheads. Launched from concealed silos, mobile vehicles, or submarines, missiles were difficult to counter before or during flight.

During the Cold War, American interest lay in developing the nuclear deterrent as a less costly option than matching the huge conventional force retained by the Soviet Union at little below wartime strength. It was also believed that the threat of possible nuclear escalation would deter 'small wars' of the Korean type, which American analysts believed would otherwise be manipulated behind the scenes by the Soviet Union, using its clients or others, but able itself to remain isolated from direct impact of the fighting. The Americans considered these a likely Soviet response in the 'rimland glacis' in an effort to break out of the United States' containment in the Mackinerist strategy of the time (Fig. 5.6). As the Soviet Union came to match American nuclear capability, America itself began to feel vulnerable to attack or nuclear blackmail, so, to deter any Soviet adventure from possibly escalating into 'general war', a doctrine of 'massive retaliation' was formulated. It was hoped this might be enough to make possible elimination of expensive garrisons in or of subsidies to volatile countries in the rimland around the Soviet heartland, as well as to reduce chances of embarrassing diplomatic incidents. The Americans believed the doctrine made war so unthinkable that it ceased to be a viable instrument of policy, making nuclear weapons a purely deterrent arm, even though it still allowed a threat of war as a policy option in diplomacy or as an ultimate sanction if all else were to fail. Deterrence only works, however, if it is credible and seen to be no bluff, so circumstances grave enough for resort to nuclear weapons had to be publicly defined and the size and capability of these weapons advertised. Secrecy and surprise, once defence keys, no longer had a role. It was also important to get the full deterrent effect by making the missile launchers invulnerable to any pre-emptive strike, with the 'first strike' backed by a retaliatory backlash 'second strike' capability, for which submarine-launched missiles were peculiarly suited.

Such 'massive retaliation' was after a while seen as an inflexible doctrine, not scaled to respond to incidents according to their seriousness, and liable to be unleashed for what turned out to be only a modest crisis. It was thus modified into 'graduated deterrence' or 'flexible response', with escalation in steps as the crisis unfolded, so that emphasis shifted back partially to conventional forces. With Soviet missiles able to penetrate America itself, there was good reason for the United States to keep any conflict well below the nuclear threshold. The unresolvable question remained whether nuclear escalation would happen almost immediately in a crisis as military necessity surmounted political caution: the dilemma has been summed up as 'initial use of nuclear weapons must be timed as late as possible but as early as necessary'!

Scared by the brinkmanship of the Cuban crisis in 1962, the super-powers initiated efforts towards *détente*, so that a deterrent-based coexistence could be renamed peace! As the cost of strategic defence systems and other armaments escalated, there was a powerful economic incentive to seek at least a balance of forces, even if still capable of 'mutually assured destruction'. Moreover, satellite imagery began to provide reliable surveillance to verify compliance with any agreement. Interpretations differed in the perceptions of the super-powers: what one saw as a move towards 'balance' was viewed by the other as a renewed threat of imbalance: only in the late 1980s has some progress appeared.

Despite preoccupation with dilemmas of nuclear defence, provision of conventional defence forces remains essential. The importance of air superiority has been demonstrated in every 'small war' since 1945, while the value of air transport for supplies was illustrated by the Berlin Blockade in 1948. Airpower has further reduced the value of interior lines and made concealment of intent ever more difficult, and it has become an important component in seapower, which remains particularly significant, with nuclear-powered ballistic missile-carrying submarines an exceptionally important postwar addition. Even an essentially 'continental' power like the Soviet Union has realized the value of seapower, developing large naval forces during the 1960s as a counter to its containment by American grand strategy and as a key factor for extending Soviet influence beyond the 'World Island'. Soviet fleets now cast an ever present shadow across the West's seaborne supply routes. The Soviet naval weakness is that approach to all its own naval bases must pass potentially dangerous 'choke points' and most of its ports are hard to keep operational through the winter. The United States has emerged as the greatest sea power, capitalizing on wartime experience in the Pacific, and its large, specially structured fleets are stationed in strategically crucial waters, ready for quick deployment.

Modified blitzkrieg of 1943-5 remains the tactical basis of land warfare, though now more mobile and destructive. The unreal theory that tactical nuclear weapons could be used on the battlefield without escalating nuclear engagement was quietly dropped. The insurmountable defence problem of Western Europe remains geographical — the limited space for strategic and tactical manoeuvre and the several narrow territorial waists. Space has become more critical as the tactical offensive has been worked into the doctrine of strategic defence, where the older forward-defence concept, with little manoeuvre and mobility, has given ground to a concept involving possible ceding of considerable ground, an idea unpalatable to the West Germans whose territory forms NATO's forward-defence zone.

Partisan and resistance groups in Occupied Europe between 1939 and 1945 provided models for guerrilla insurgency characterizing most subsequent 'small wars'. With all the inherent risks of escalation, direct assault upon an enemy is now less likely than attempted destabilization of a government before seizure of power. Insurgency has frequently been fuelled secretly by one superpower against a government supported by the other, while a new phenomenon has been terrorism pursued by political or religious fanatics, often covertly encouraged by ulterior interests. The appropriate response is generally a combination of political and economic measures to isolate the insurgents and police or military action to winkle them out. From key points small mobile forces fan out over ground prepared by undercover agents, winning popular support and preventing the insurgents from trading space for time. Small numbers of insurgents can tie down large government forces, often exceeding one to ten. The difficulties of combating insurgency or terrorism are notably great in liberal democracies, where severe restrictions on personal freedom and mobility are unacceptable. In totalitarian regimes internal passports and control of residence and movement have long been imposed. The spectre of security management becoming integral in urban and regional planning does not seem far away.

The British successfully combined political, military, and economic measures against Communist insurgency in Malaya, whereas the French and Americans were less successful in the more difficult geographical setting of Indo-China, and Russia appears to have achieved little in Afghanistan. This kind of war is not new, for it was successfully used by the Romans and Normans. In the middle ages ruthless 'harrying', punitive and brutal, applying a 'scorched earth' policy was frequent, just as 'plantation' by fortified settlements of loyal supporters was also used, like the Roman colonies of veterans or the Plantagenet 'bastides' of North Wales. Partisan success often lies in the nature of the terrain, well seen in the mountains of Yugoslavia, the Russian forests, or the Massif Central of France in the 1939–45 War.

Changes in the conduct and technology of war, affecting the value of space and terrain as well as human and other resources, have exercised an influence on the way states have organized their territory and how they have viewed the value of its component parts besides its resources for national defence. Surprisingly little attention has been given to the relationships between political-geographical issues and the military-geographical problems of defence.

6

Territorial administration

For convenience of internal government, states commonly divide their territory into administrative districts, often in a tiered hierarchy, each level having clearly defined functions and responsibilities. Apart from tiny countries, even the most centralized state would find a single territorial administration for internal government unacceptably complex. Division into local-government districts devolves a range of everyday functions and responsibilities, allowing local decision-making to reflect local needs, and possibly involving local people in the process of government. How far local government can influence central-government policy depends on the degree of devolution and decentralization, which may go furthest in a federal system (e.g. Switzerland), though others may concede a significant level where there is the will to do so (e.g. United Kingdom). Local-government hierarchies, even where there is a federalist element, may nevertheless in reality leave a high degree of control in central hands (e.g. Soviet Union). Liberal-democratic local government usually reflects a strong historical dimension built on powerful loyalties and identities, with systems primarily concerned to provide services geared to local needs and wishes. The centrally planned economies and totalitarian regimes put more emphasis on territorial-administrative systems serving spatial management of society and the economy and usually depart radically from any historical dimension.

Over time, all territorial-administrative systems need revision to keep reasonably in balance with changes in the economic, social, and population geography of the country. This adjustment is most difficult in liberal democracies, where strong local allegiances develop and local vested interests are averse to modifications liable to weaken them. In contrast, the centrally planned economies often make frequent adjustments in the number and boundaries of administrative units, mostly in response to economic changes. Most resistant to change are the component *states* of a federal system that enjoy a measure of self-government and considerable popular sentiment, well illustrated by ill-fared attempts at change in the United States or in Federal Germany. In Nigeria efforts to change patently unsuitable federal boundaries resulted in civil commotion.

Distinction can be made between territorial-administrative units whose boundaries reflect a strong historical dimension and those imposed with little or no reference to any earlier system. To the first belong the English counties, many of whose boundaries may be traced back to pre-Norman times; to the second the Soviet *oblasti* that entirely replaced the pre-revolutionary pattern. France occupies an intermediate position, for the Revolution swept away the old system and substituted *départements* whose boundaries have since been little altered. West Germany has units both of historical standing and ones created after 1945.

An important decision in any territorial-administrative reform is how parochial loyalties should be treated. Encouragement facilitates formation of local pressure groups or even dreams of regional autonomy: discouragement is usually aimed at focusing loyalty on the nation. The drastic reform of local government in Revolutionary France was partly aimed to create a strong national identity and loyalty to the Republic ('one and indivisible', according to Robespierre). Local government development in the nineteenth century in the United States was concerned with creating democratic government at all levels of the federal structure, emphasizing public participation. A remarkable degree of local management of affairs existed in England until late last century, since when central government has tended, even if sporadically, to intrude increasingly into local government.

Planners of territorial-administrative systems need to consider how far component districts should reflect the spatial pattern of the national economy. Should industrial districts be under one administration or can they be divided among several? This became an issue in territorial reform in the Weimar Republic, with disagreement over a proposal to bring the Ruhr coalfield under one authority rather than divide it between two provinces and three *Regierungsbezirke*. The fear was that a 'Ruhr Coal Province' would unduly enhance the power of heavy industry and would be too dependent on a limited economic base. The proposal was finally reduced to an overall planning authority, co-ordinating the infrastructural needs of the coalfield. Even in the Soviet Union complete identity between administrative boundaries and the broad economic regions proposed by Kalinin's Electrification

Commission (*GOELRO*) of 1921-2 faltered because of Lenin's insistence on the sanctity of union republican boundaries and meant, for example, that a unified Donbass could not be achieved.

The boundaries of the different tiers of local government invariably 'nest' — no English parish, for example, lies across two counties. Local government districts in Britain nevertheless have to deal with a range of 'areas' defined for administrative convenience by public utilities and central government agencies, whose boundaries have unfortunately not always shown correspondence with the higher echelons of local government.

A dilemma in any reform is the time-scale envisaged for the system's life before another reform appears desirable. Should the attempt be to design districts viable for, say, 50 years or more, or should 'life-expectancy' be merely 15-20 years or less? The infrastructure of local government is expensive to establish, so a reasonable life is important. In Western Europe change has been slow and infrequent, usually with only modest alterations rather than fundamental recasting, allowing development of community. The rapid change in economic and population geography has nevertheless witnessed frequent and quite radical reform in the Soviet Union.

Should the structure of the system and size of units be applied uniformly across the national territory or allowance made for exceptional areas? Local government structure and areas are commonly not uniform: before the major reform of 1974 in Britain, differences existed between England and Scotland which, if anything, were further accentuated by the reform. In West Germany small differences between the constituent *Länder* date back to before 1871. Of course, whatever the system, it is common to give some districts specially favourable economic treatment (e.g. development areas in Britain, *Zonenrandgebiet* in West Germany).

Fair reflection of ethnic minorities in the territorial-administrative structure faces several dilemmas. First, minorities are unlikely to be given special treatment if their perceived loyalty is suspect — in such an instance, the system is more likely to be designed to divide and weaken them. Much depends on whether central government is prepared to let them cultivate their identity or seeks to integrate them completely. Second, they are unlikely to have special districts unless they form really substantial elements in population. Third, even with special districts, how far any measure of autonomy is conceded, other than in culture and language, depends on central government's fears of ultimate secessionist pressures. In the Soviet Union 'national' territories exist at all levels of the territorial-administrative hierarchy and Lenin stressed that their boundaries should be inviolable, a view still respected if recently questioned. It is worth noting that the titular group does not always form a majority in its district, even at republic level (e.g. in Kazakhstan). In India territorial-administrative reorganization of the confusingly complex pattern has sought to erect first-order administrative units based where possible on linguistic patterns, a policy apparently successful in the polyglot environment. In mainland China, in addition to twenty-one provinces and two special municipalities (Peking and Shanghai), there are five autonomous regions for non-Han peoples, containing, however, only 6 per cent of the population but covering 40 per cent of national territory. Small ethnic groups have their own districts at quite low levels of local government (*chou* and *hsien*), but everywhere in a highly centralized state ethnic autonomy is restricted to cultural matters. In East Germany, despite encouragement to the Lusation Sorbs to foster their identity, the government has not matched it by special territorial arrangements, so they live scattered between the *Bezirke* Cottbus and Dresden. Extensive changes in the territorial-administrative structure of Rumania since 1945 have several times altered the status and extent of the Magyar districts of Transylvania, where a modest level of autonomy has been enjoyed.

In structuring a territorial-administrative system it must be decided whether it will comprise a few large or many small districts. Small districts may better reflect socio-economic groupings, but larger ones are preferred where emphasis is on efficient services through economies of scale, as well as for planning and administrative efficiency. Where such units form a basis for national planning, size is an important consideration, since the need is to mirror accurately the true diversity and yet not divide into an unmanageable number, a source of much debate in the Soviet Union. In the liberal-democratic system the important consideration is how the size of districts will affect formation of political pressure groups and their implications for central government, as well as how it may also affect public willingness to participate in local government — small units probably encourage local interest, while large units, with more distant seats of administration and wider issues to settle, may tend to discourage interest.

ELECTORAL GEOGRAPHY

A special aspect of the internal territorial division of countries is the

133

arrangement of voting districts for popular representation, most especially for central government. Such boundaries allow some scope for manipulation, affecting the voting outcome in any election by positioning them in relation to various socio-economic groupings or known lobbies. A key decision is often whether boundaries should consolidate or divide social and economic groups or religious and ethnic affiliations. Electoral geography is really only a meaningful study where there is free expression of party political preferences in a multi-party system in which no one party is in a superior position to any other. It has little significance in countries with a one-party system or where parties recognize the superiority of the government party or where all are subjoined into a national front.

Voting systems vary considerably from one country to another, some remarkably complex to aid truly proportional representation, others quite simple. In Britain in each constituency the winner elected to parliament is the candidate with the most votes, but in a three-cornered fight, for example, this candidate might only poll 34 per cent of the votes, thus not representing the views of the remaining two-thirds. In France a two-stage voting system means that only candidates who poll more votes than all the others together are elected in the first round: otherwise a second ballot is held between the two top candidates. The system was developed to deal fairly with the great number of political parties traditionally existing in France, as a safeguard against election of representatives in large numbers on minority votes. Some countries have schemes whereby voters express preferences rather than cast a single vote for a candidate (or candidates in multiple representation): this method is used in the Australian states, in Malta, and in Ireland among other countries. Other systems based on party lists (though possibly allowing discrimination between a party's candidates) have been common in Europe. Some countries have used mixed systems, primarily to achieve proportional represent-ation with greater certainty, notably in Denmark, Iceland, and West Germany (which employs a nationally based balancing operation of 'topping up' seats). No system seems beyond quirks and idiosyncrasies in operation.

A carefully demarcated pattern of constituencies is needed across the national territory, so that each should contain within close limits a similar number of electors, apart from special circum-stances. Because the pattern of population distribution changes, constituency boundaries need to be reviewed at regular intervals and, if necessary, modified. Fairness and impartiality are not always applied, with boundaries consciously drawn to produce bias in the voting pattern to the advantage of a particular party.

In Britain the Industrial Revolution, with the rapid growth of population, brought pressure to expand the franchise and create a more equitable representation. In the eighteenth century the franchise was vested powerfully in the propertied classes of the older towns and in rural landowners. The growing number of large industrial towns in the North and Midlands sent no representatives to Westminster, while some small places ('rotten boroughs') mostly in the South and South-west, with a hundred or so voters each, might even have two members (in 1785 Pitt had tried unsuccessfully to dissolve thirty-six of the weakest). In 1800 only 3 per cent of the adult male population possessed a vote and ten counties south of a line from Bristol to the lower Thames elected two-fifths of the members of the House of Commons. The Reform Act of 1832 modified rather than swept away anomalies and abuses. Though the new industrial towns remained underrepresented, at least forty-one large towns got representation for the first time (including Manchester, Birmingham, and Bradford). Despite a large increase in the electorate (by over 1,000 per cent in Scotland and even 21 per cent in Ireland), England, with 54 per cent of the Kingdom's population, still returned 71 per cent of the members of parliament.

The franchise was again extended by acts of 1867 and 1884 and a more equitable distribution of seats came in the act of 1885. Even so, less than a third of the adult population possessed a vote before the Representation of the People Act in 1918 gave all adult males the vote, plus a limited female suffrage, until full adult suffrage was achieved in 1923. The franchise for universities and business premises was not swept away until 1948.

Glaring inequalities between electoral numbers in constituencies (often over 10:1) lingered into this century, but this has been surmounted by periodical reviews every ten to fifteen years to redefine boundaries to ensure as equal as possible divisions. Knowing the number of parliamentary seats, the Boundary Commission can divide the electorate by that number to gain an electoral quota, indicating how many electors there should be to each member of parliament and consequently how many constituencies. Some variation from the electoral quota in constituencies is accepted, so it has been usual for Scotland and Wales to have more seats than their strict entitlement, whereas England (unlike the early nineteenth century) is underrepresented. Overrepresentation of Scotland and Wales is possibly to avert stimulating nationalism if seats were lost, while Ulster's underrepresentation for many years was offset by the Stormont

Table 6.1 United Kingdom parliamentary representation, 1984

Electoral quota: 66,129 electors per constituency

	Entitled	Actual		
England	541	523	Underrepresentation	18
Wales	33	38	Overrepresentation	5
Scotland	60	72	''	12
Northern Ireland	16	17	''	1

*numbers rounded
Source: Statesman's yearbook, 1985–1986

regional parliament's dealing with some matters otherwise the concern of Westminster.

In thinly settled rural areas, like highland Scotland or upland Wales, or in densely crowded urban areas, a constituency may have an electorate as much as 25 per cent above or below the electoral quota at the major regional level. Where people are thinly scattered over a wide area, to gather together the electoral quota in full would mean a really huge constituency with unwarrantable amounts of travelling for campaigning or meeting constituents. In urban areas high concentrations of constituents allow extremely easy access in a very small space, so the electorate can be acceptably above the quota. On this basis in 1987 the average constituency size was 61,364 but ranged from 98,694 in the Isle of Wight down to 23,507 in the Scottish Western Isles. The distortion overvalues rural votes and undervalues urban ones, with obvious political implications. Such a pattern has worked against the Communists in France, whose strength has lain in industrial towns, while before 1945 in Central Europe rural votes were given preference to protect agrarian interests, just as in Ireland preference has been in favour of small family farms in the Celtic West, seen as the core of the nation. In Britain electoral disproportions are seemingly primarily for administrative convenience, with constituency boundaries coinciding as far as possible with local government ones, reflecting local community interest but also meaning that few constituencies comprise both rural and major urban areas.

A special distortion of electoral boundaries is where they are drawn to maximize potential support for one party, creating a 'safe seat' at the opposition's expense. This 'gerrymandering' (after Governor Gerry of Massachusetts whose Boston districting law of 1812 produced a 'salamander-shaped' district favouring his own party) can be used *either* to concentrate opposition voters into a few districts (where they win easily but nowhere else) *or* to break up concentrations of voters representing specific parties or interests, so their votes are dispersed and thus wasted. To achieve either, voting districts usually become grotesque shapes or are even broken into several 'islands'. Despite considerable legislation in the American Supreme Court, this undesirable and undemocratic practice has not yet been eradicated.

LOCAL-GOVERNMENT STRUCTURES — SELECTED EXAMPLES

United Kingdom

Anglo-Saxons and Normans modelled local government in England and influenced Scotland, Wales, and Ireland. The early territorial pattern was primarily for administration of justice and preservation of law and order, but also for tax collection and organizing defence. During the tenth century, the *shire* (Old English : administration) emerged, first in the English Midlands, each based on an important castle, with its limits in inaccessible forest or marsh. The earl or count as governor for the king generally sought to encompass his own estates within his authority, accounting for the shape of many of the earlier shires. Soon the administration passed to the sheriff ('shire-reeve') as the king's chief agent, though his power was to be curbed from the fourteenth century. The lowest level of local government was the Anglo-Saxon *tun* (the Domesday *vill*), usually coincident with a manor responsible for law and order, tenancy, rents, and tithes. Considerable local government leadership came from the church whose affairs were managed by elected wardens. As parishes so often coincided with the general extent of tuns (townships) or manors, local government generally passed to them. In Norman times, to provide a reasonable lower limit of direct central government control, the old tithings of freemen were organized into *hundreds* each supervised by a bailiff, direct representative of the sheriff, with responsibilities for maintaining roads and bridges and for the poor and vagrants. Over much of nineteenth-century England, the Poor Law Union boundaries coincided with the otherwise well-nigh defunct hundreds.

Special royal favour was shown to districts important for the security of the realm, with privileges (mostly in legal and military matters) granted to valuable duchies (like Lancashire and Cornwall) or to strategically significant counties palatine (like Chester and Durham). Other 'liberties' were also granted, usually for service to the Crown, like the Soke of Peterborough and the Isle of Ely, both

137

later administrative counties. From the early middle ages, through their growing wealth and influence, towns won considerable rights to manage their own affairs, to raise money and to hold markets and fairs, conceded to them in charters. A charter usually constrained the rights of the sheriff and even of the king over the town, its mayor, and corporation.

In Scotland Anglo-Norman influence spread the county system by adapting older institutions, first in the lowlands and east coast. Some institutions of Anglo-Saxon origin, like thanages, already existed and small shires had been established. True sheriffdoms appeared by the twelfth century, though away from the lowlands and east coast they came late — Caithness did not become one until 1503 and Orkney only in 1530. Some sheriffdoms in the east and northeast were probably modelled around ancient Pictish provinces. Many early small sheriffdoms soon merged into larger ones, but boundaries were reasonably fixed by the fourteenth century, with little subsequent change. In the north and northeast the territorial pattern was complicated by numerous small enclaves, some surviving until last century, despite considerable reorganization by James VI. Several chartered burghs existed before the twelfth century, when more were instituted. The thirteenth-century English conquest of Wales absorbed much of it into Crown lands but was subsequently followed by institution of a county system on the preferred English model, finalized in 1536. That Act of Union swept away the marcher lordships, powerful families jurisdictionally outside the normal framework of English government and absorbed them into the thirteen counties.

Under the Tudors expanding commerce and the growth of towns, combined with collapse of care of the poor (previously provided by the now defunct monasteries and weakened church), made a new pattern of local government desirable. From 1555 to 1601 the parish was given added duties, including administering the poor law, while the power and responsibility of the justices of the peace were increased; as these were Crown appointments, the power of central government was correspondingly increased. Local autonomy grew during the Civil War and in 1688 no central-government interference in local affairs became a key principle, remaining so well into the last century. Central-government influence could be exercised, however, through the High Court but was seldom used, being expensive and ponderous. This situation worked because local government was essentially low key.

Eighteenth-century local government, even if efficiently run and despite low demands made upon it, was inadequate to provide services increasingly required in an industrializing economy and society. To cover many services (like lighting or poor relief), commissioners or corporations responsible for specific districts were provided and turnpike trusts took over main roads, though local roads remained a parish upkeep. More and more bodies were set up, whose responsibilities and areas overlapped in a nightmare of confusion until sorted out late in the nineteenth century. In 1871 the Local Government Board in England and Wales was made responsible for central government's supervision of local authorities (from 1919 the Ministry of Health), whose responsibilities were defined in an Act of 1872, and borough organization reformed in 1882. So began a slow encroachment of central-government influence over local government continuing intermittently to this day.

The county borough was created in 1888, when towns (usually with over 50,000 people) could be given status equivalent to a county, while new administrative counties were formed from the ancient geographical counties, both with popularly elected councils. In England and Wales outside London, urban and rural districts were constituted from old sanitary districts and parish organization modernized. In 1899 London was reorganized into metropolitan borough councils. Public education (1902) became the responsibility of counties and county boroughs as well as the poor law (1929).

The 1888 Act created fifty-nine county boroughs in England and Wales and another twenty were added by 1923, so that counties were alarmed at the proliferation, with fears particularly expressed for the future of Lancashire and the West Riding. In 1926 creation of county boroughs was made more difficult, lifting the population threshold to 75,000, and until 1958 only one further addition (Doncaster) was made, although extensions to boundaries, albeit mostly small, of existing boroughs continued. Changes in the counties arose principally from erosion by county boroughs, but the old geographical counties had been converted into forty-eight administrative counties, often through division. Traditional Yorkshire split between three Ridings; Lincolnshire was divided into three; Suffolk, Sussex, Cambridgeshire, and Northamptonshire each divided into two. Division rested principally on the location of quarter session courts in an ancient county, so consequently little Suffolk was divided whereas extensive Lancashire was not: such quirks of legal history hardly seemed a good criterion for division! In 1890 the Isle of Wight was separated from Hampshire and a few modest boundary changes came after a statutory review in 1929.

The act of 1958 raised the threshold population of the county borough to 100,000, with a few new ones created but some existing ones demoted, although extension to county-borough boundaries continued (especially around Birmingham). This act reduced the administrative counties to forty-five: London and Middlesex disappeared on formation of the Greater London Council (not itself a county), while Cambridgeshire and the Isle of Ely merged, as did the Soke of Peterborough and Huntingdonshire, but mergers between Leicestershire and Rutland and between Holland and Kesteven in Lincolnshire were not pursued. The boundaries of urban and rural districts had changed greatly, though the total number remained within close limits, ranging between 1,520 and 1,580 between 1894 and 1929 (of the 290 new districts created in that time, only 26 were rural ones), while by 1939 the number of districts had fallen to 1,222 (small boroughs had grown from 256 to 277), and the 1958 act further reduced the number to 1,119 (mostly through change in the West Midlands and Greater London). After 1958 six small boroughs in Shropshire and Cornwall were reduced to parish councils, though allowed to retain their mayor and corporation, being known as rural boroughs (Fig. 6.1).

A fundamental reform of this patched system was instituted in 1966, with the intent that local government should respond to the patterns of society, operate through democratic procedures, and be truly cost effective. Three working principles were enunciated: all related services to be together under unified control for efficiency and effectiveness; the size of units to be able to encompass all 'externalities' like commuting and environmental needs; local government areas to be not less than 250,000 but not more than 1,000,000 people. The upper threshold was set to maintain ready accessibility to personal services. The general model was the county borough because of its functional co-ordination, while preservation of existing and historically justified boundaries was accepted as significant as a focus of local identity.

To define new unitary areas of about 250,000 people, patterns of population change, shopping, transport, and commuting were examined and zones of indifference between major towns were identified. The city-region approach did not work well in some parts, notably Northwest and Southwest England, with much confusing overlap in the Mersey Basin and Yorkshire. The one-million-population maximum was applied in only three metropolitan areas: Merseyside, Greater Manchester, and Birmingham, each with a two-tier structure to preserve good access to personal services, with some twenty districts in all. The metropolitan units were to cover matters

Figure 6.1 United Kingdom territorial administration pre- and post-1974.

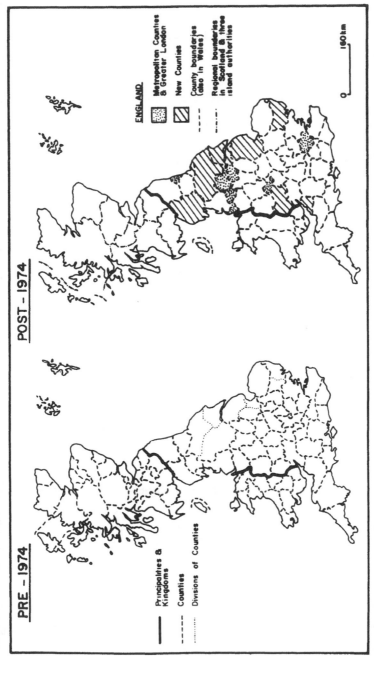

PRE – 1974

POST – 1974

ENGLAND

Metropolitan Counties & Greater London

New Counties

County boundaries (also in Wales)

Regional boundaries in Scotland & three island authorities

Principalities & Kingdoms

Counties

Divisions of Counties

0   160 km

The effect of local government reforms has been spatially most striking in Scotland and Wales

Figure 6.2 Royal Commission Proposals on Local Government in England, 1969.

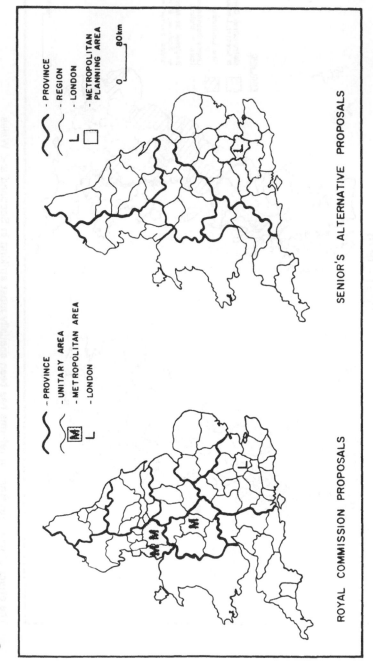

ROYAL COMMISSION PROPOSALS

SENIOR'S ALTERNATIVE PROPOSALS

~ – PROVINCE
M – UNITARY AREA
   – METROPOLITAN AREA
L – LONDON

~ – PROVINCE
~ – REGION
L – LONDON
□ – METROPOLITAN
     PLANNING AREA

0   80km

The Commission's proposals were accompanied by a dissenting view. Adapted from *Royal Commission on Local Government in England*, HMSO Cmd 4040, London 1969 like transport, planning, housing, and environmental issues, the districts serving personal services. Apart from the three metropolitan units, there were to be fifty-six unitary authorities (three with over a million people each), some in essence counties, others city-regions in entirely new boundaries. Additionally eight provinces (corresponding remarkably to the existing regional planning councils), rooted in local government, would manage broad economic, investment, and land use in planning and development of the authorities within them (Fig. 6.2) A 'memorandum of dissent' to these proposals argued for the size of units to be determined solely by the way society was organized and not by arbitrarily applied thresholds and population maxima. It saw no part for historical boundaries, proposing a division of England into 35 provinces comprising 148 districts (50 districts had populations greater than the smallest province).

The new government returned in 1970 changed the approach, though the general objectives and principles remained similar, choosing to retain where possible the historical county boundaries. Most services would be provided by counties of about 250,000 people, but local matters handled by districts of about 75,000 population. Large metropolitan counties, divided into districts of about 250,000 people, would operate as originally proposed. Wherever possible, existing counties would be preserved, with those below the threshold amalgamated and a few new ones created where population growth had strikingly altered the traditional pattern. Boundary changes would seek to eradicate anomalies or barriers to effective administration. In the end England was divided into 39 non-metropolitan (in all about 296 districts) and six metropolitan counties (containing 36 districts). In Wales 8 counties, divided into 37 districts, replaced the previous 13, and community councils like the English parish councils were created. Scotland, subject of a separate examination, saw the 33 civil counties and 4 counties of cities replaced by eight large regions and three island councils, in all some 53 districts. Strathclyde region contains 46 per cent of the national population, while Highland region covers one-third of the national area but contains hardly 4 per cent of total population.

France

The spatial pattern and structure of local government in France have changed little since the Revolution, with most change coming in the 1960s. The British idea of local self-government found little response

Figure 6.3 France — territorial administration

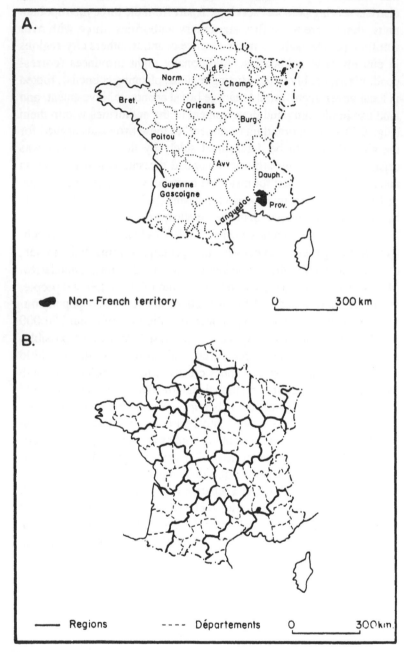

Map A shows the old provinces before 1789 and Map B the *départements* and *régions* in the early 1980s. Compiled from various sources

in France, with neither a tradition of nor a clamour for local management. Permanent officials appointed by and responsible to central
government conduct local administration, with the emphasis on a
system to assert national unity in a uniform state run by obedient local
agents dependent for their power on Paris.

Under the *ancien régime* France was divided into thirty large and
seven small military governments termed provinces, which in their
government and legal organization showed considerable differences,
a confusing and inconsistent pattern where, from the days of Louis
XIV, central control had been increasing. The Revolution swept away
the old provinces, with their strong local loyalties, and instituted 83
départements, territorial-administrative units of remarkably equal size,
conceived principally for administrative convenience and strong
central control. Boundaries cut across and destroyed the fabric of the
old provinces, with *départements* named after physical features and
easily supervised from the town chosen as the seat of the prefect and
advisory council, appointed from Paris. Direct administration included
supervision of subordinate territorial-administrative units (notably the
communes) which contributed elected members to a general council
in each *département*, effectively to approve matters submitted by the
prefect.

Of the subordinate levels, the *arrondissement* and *cantons* have
purely administrative functions, so local government is really conducted by the *commune*, nowadays numbering over 36,000 in the 96
départements, varying in area and population (over 31,500 have less
than 1,500 people and over 16,000 fewer than 300, but 230 exceed
30,000 people). The mayor of the *commune* is an agent of central
government, ensuring execution of state laws and decrees, supervising
police, public works, and revenue.

The 83 *départements* of 1790 later increased by territorial gains
and between the world wars numbered 90, but after 1945 increased
to the present 96, pre-eminently through reorganization around Paris,
the focus of population growth. Nineteenth-century France, unlike
Britain and Germany, did not experience massive population shifts,
making territorial reorganization less pressing. The main change was
in the 1960s, by creation of additional *départements* and modification of *commune* boundaries, but also in some decentralization under
growing dissatisfaction with the overdominance of Paris in national
life ('Paris et le désert français'). To stimulate regional development
and assuage local pressures for identity, 22 *régions de programme*
have been created for planning and development as well as for
budgetary policy, named after the long-ignored provinces, with

145

executive power since 1982 under the president of their directly elected regional councils.

Germany

The German lands were a patchwork of small, independently minded territories, loosely associated together in the Holy Roman Empire until 1806. Even the largest were modest in area, population, and wealth. In the sixteenth century the Emperor had tried to get order into the administration by grouping territories together in six (later ten) 'imperial circles', but with little effect on the behaviour of the particularist states, a *Kleinstaaterei* of some 360 sovereign territories slimmed down to 35 states and 4 free cities in 1815.

Prussia, the largest and most dynamic state, emerged as a considerable power in the eighteenth century, creating a need for effective internal management to raise taxes from modest resources to maintain the large army. The foundation was laid when the administration of the royal estates, the war commissariat, and the taxation authorities were brought together in a *Generaldirektorium*, with eventually (1786) some fourteen *Kammerbezirke* for management convenience. The boundaries of these units corresponded roughly with the geographical and historical provinces, although some of the larger ones were divided into two or more *Bezirke*. The *Generaldirektorium* itself was divided into four sections, each responsible for two to four *Kammerbezirke*. Education, church matters and the judiciary were retained by the old provinces.

To adjust to the strains of the Napoleonic period, a reform of 1808 laid the foundation of the modern administrative map. The *Kammerbezirke*, in modified boundaries, formed a basis for the modern *Regierungsbezirke*, large administrative units without self-government, and their number increased to twenty-five through territorial growth in 1815. Each averaged about 400,000 inhabitants. A new element was the 'province', often bearing a marked relation to its historical predecessor, and the upper echelon of local self-government. Below lay the *Kreise*, towns and *Gemeinden* (communes), with a very limited franchise, though through the nineteenth century a slow move to greater local responsibility ensued, most successful in municipal government. Even so, until 1928, in parts of Prussia east of the Elbe the Junker landlord of the *Gutsbezirk* ran administration and jurisdiction effectively without popular participation.

146

A modified Prussian model was followed in other large states, notably Saxony, Baden, Württemberg, and Bavaria. The latter was, for instance, divided into eight *Kreisregierungen*, which in contrast to the Prussian *Regierungsbezirke* had their own elected self-government, though usually the terminology rather than the function varied. Even within Prussia, there were differences, where, like parts of South Germany, some areas had *Gemeinden* grouped into an *Amt*. Some Prussian towns had the constitution enacted in the 1808 reform, others (notably in the Rhineland and Westphalia) had their constitutional basis inherited from Napoleonic France, as adopted in Baden, Saxony, and Thüringen. Prussian leadership after 1871 made little difference to the territorial-administrative pattern as states continued in their own manner. Though the *Reich* had the same monarch and seat of government as Prussia, a substantial duality existed, for the states resisted creation of *Reich* institutions which might have changed their own internal systems, although there was nevertheless penetration, particularly in taxation and fiscal matters.

Despite the upsurge in industry and towns after 1871, administrative change was surprisingly modest, much centred on the Ruhr coalfield, industrial central Germany, and Silesia. Between 1873 and 1912 a collection of new independent boroughs (*Stadtkreise*) emerged in the Ruhr and many towns had boundary extensions, following the example of Berlin, which, after the union of five towns in 1710, had had its boundaries extended in 1841 and four times between 1881 and 1915. After 1919, under the Weimar Constitution, the monarchial states became 'free states' (usually termed *Land*/pl. *Länder*), but their internal division was not altered, though the franchise widened to universal suffrage (Fig. 6.4). Even so, local authorities were soon robbed of their independence as Germany's financial difficulties multiplied. The new Republic faced a clamour for a complete revision of the provincial and *Länder* boundaries to replace historical anomalies and the complex spatial pattern by units reflecting the social and economic alignment of the new age, yet preserving the strong regional identities of the German people. Numerous proposals, some well argued, others hasty or emotional, were made, but radical change was not achieved, notably through a reluctance to break up Prussia, fearing that that would further weaken post-Versailles Germany, while few groups were prepared to merge their identity and freedom of action into a broader spectrum. The situation was not helped by separatist movements in the Rhineland fed from outside.

Nevertheless, change was made, extending into the Nazi period, most notably the creation of Greater Berlin (1920), though a similar

Figure 6.4 Territorial administration in interwar Germany.

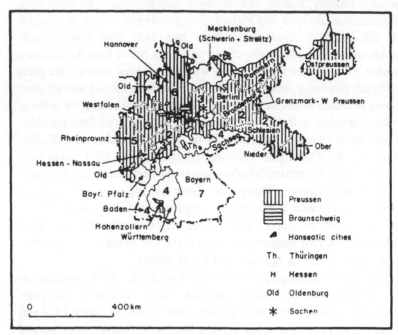

Numbers represent the number of *Regierungsbezirke* in each state

move for the Ruhr did not succeed (p. 131). City boundaries were widely revised, especially in the Ruhr, around Frankfurt, and in Greater Hamburg (1937). A main achievement was elimination of the patchwork of small territories consolidated in *Land* Thüringen (1920) and some small units merged with Prussia (Waldeck, Pyrmont etc.), while the two Mecklenburgs merged (1937) and scattered Oldenburg territories were 'rationalized'. Despite this, even by 1945, many hundreds of tiny enclaves still remained.

The National Socialists established an elaborate territorial party organization, headed by the *Gau* and with the base in the party agent (*Blockwart*), serving a group of dwellings. Though this was not part of the 'state' administration, the all embracing nature of the party gave it a powerful influence, for well before 1939 most *Oberpräsidenten* of the *Länder* were also *Gauleiter*. Despite Hitler's reassurances local government and party organization would ultimately have fused, seen in the *Reichsgaue* set up in territories incorporated into Greater Germany. The *Gaue* were in many parts coterminous with existing *Länder*, though this correspondence was least in the west and centre, where high population density was reflected in a really small *Gaue*.

148

Figure 6.5 Territorial administration in the two German states in the 1970s.

The federal structure of *Länder* in the West contrasts with the centralized system of *Bezirke* in the East.

The initial intention of the Allied powers in 1945 was to create a German federation. Though Prussia was completely dissolved, the historical *Länder* were retained little altered, but there were also some

149

entirely new creations. The subordinate units within them changed little, with elective local government modelled on the Weimar Constitution. The new West German Republic retained the federal structure, though its Basic Law provides for territorial change, but the *Länder* have altered little, despite numerous proposals either for a return to vanished prewar boundaries or for radical realignment to cope with postwar conditions (Fig. 6.5). The great postwar changes in the economy and society have been matched in local government within the *Länder*, as the widening range of services expected from and supplied by local authorities, besides the growth in the size of communities, have resulted in a rationalization of the responsibilities and size of *Kreise* and *Gemeinden*, usually with fewer and larger units. Around 1960 the Federal Republic contained 33 *Regierungsbezirke*, 139 *Stadtkreise*, and 425 *Landkreise* containing 24,525 *Gemeinden*: by the mid-1980s, there were only 25 *Regierungsbezirke*, 89 *Stadtkreise*, 236 *Landkreise*, and merely 8,471 *Gemeinden*.

The German Democratic Republic (a centrally planned state) in the early 1950s replaced its *Länder* by fourteen *Bezirke*, modelled on Soviet experience. *Kreise* and *Gemeinden* were retained, though usually enlarged and fewer in number. The *Bezirke* are much concerned with administration of planning objectives, each with a distinct 'keynote' in its economic geography, for example, all the Baltic coast's maritime interests are encompassed by the *Bezirk* Rostock (Fig. 6.5).

The Soviet Union

Peter the Great in 1708 introduced a territorial administration that was to last little changed for two hundred years, with territorial growth adding new units piecemeal. Apart from the creation of numerous small districts in the reforms of 1861, no major revision came until the Revolution in 1917. The top of the three-tier system was the *guberniya* (or *oblast* in Cossack lands), numbering 99 early this century, run on broad lines mapped out by St Petersburg. Below these lay 600 *uyezdy* and at the bottom some 15,000 *volosti*, each covering several villages. There were also a few special districts in the Caucasus and the Central Asian khanates of Khiva and Bokhara. This system mostly ignored social and economic geography as well as the ethnic pattern of the empire.

After the Revolution a commission drew up a plan for radical reform, emphasising the links between the territorial-administrative

150

system and economic regionalization as well as cognizance of ethnic nationalities. Lenin's influence gave the 'national principle' precedence over any other criteria, allowing several levels of autonomy for national territorial-administrative units, and nationalities could move from one level to another depending on circumstances. The largest and most advanced nationalities form the top level, the union republics (currently 15), which under the Soviet Constitution may secede from the Union and must consequently have a boundary with a non-Soviet country (Fig. 6.6). This condition means nationalities living in the interior (however large or advanced) may never rise above the more restrictive autonomous republic (20 in all, 16 in the Russian Republic). Small and less advanced peoples may have their own autonomous *oblast* (presently eight) or a national *okrug* (10).

The original territorial administration in the larger union republics was a three-tier structure, *oblast* (or *kray* for large territories in the less developed parts), *okrug*, and *rayon*. The smaller union republics are usually divided only into *rayony*, though the three Baltic republics briefly contained *oblasti*. The *okrug* was discontinued during the 1930s through a desire for more direct management of the *rayon* during rapid economic development. The original *oblast* (*kray*) was a vast unit, one, for example, covering the whole Ural. With discontinuance of the *okrug*, the *oblast* (*kray*) and the *rayon* were generally reduced in size and multiplied in number for more direct management, the degree of subdivision being related to economic development and settlement growth. This policy was pursued consistently until the 1960s, with frequent revision of boundaries, part of the price paid for linking territorial administration to economic planning and development, but subsequently an element of uncertainty has emerged, first in experiments with territorial organization in the problematic Kazakh republic, and second in murmurings of doubt over the continuing relevance of the 'national principle'.

The *rayon* comprises collective farms, workers' settlements, and small towns. Bigger towns are ranked according to which higher level of administration (*oblast*, republic) has jurisdiction over them. To qualify for urban status, a settlement much achieve a threshold value of adult population and industrial employment. The *oblast* typifies the principle that economic-planning needs predominate in territorial administration, termed in Soviet sources 'economic-administrative units'. The *oblast* seat is invariably an industrial town, a 'proletarian centre', regarded as a focal point for dissemination of Communist ideology. Its territory is in most respects its economic service area,

Figure 6.6 USSR — territorial administration.

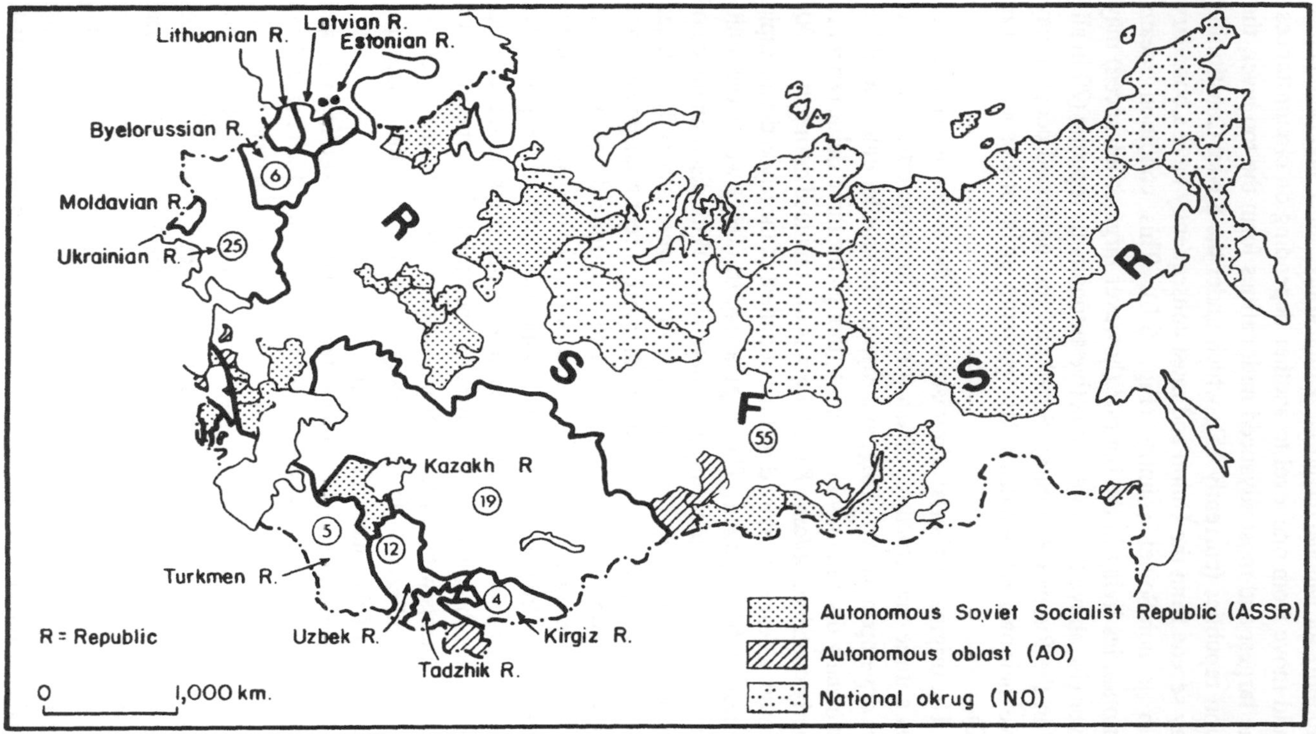

Lithuanian R.
Latvian R.
Estonian R.

Byelorussian R.

Moldavian R.

Ukrainian R.

25

6

R

S

R

S

F

55

Kazakh R.

19

Uzbek R.

12

Turkmen R.

5

Kirgiz R.

Tadzhik R.

4

Autonomous Soviet Socialist Republic (ASSR)

Autonomous oblast (AO)

National okrug (NO)

R = Republic

0   1,000 km.

The numbers in circles represent the number of *oblasti* in each union republic where applicable from which a 'complex economy' is developed, following any specialization on an 'all-union principle' and pursuing a high level of self-sufficiency, chiefly to reduce the burden on the national transport system.

There are also large economic regions (*makrorayony*) modelled on *oblast* and republic boundaries, used for general planning, fiscal, and statistical purposes by the State Planning Commission (*Gosplan*). Twenty-one such units proposed in 1921–2 by Kalinin's State Electrification Commission were rejected because they did not respect the national principle. Over a decade later an amended *Gosplan* system was instituted, whose boundaries have been modified several times, with the number of regions ranging from 13 to 19. These Gosplan units have become standard for regional description in the Soviet Union (Fig. 6.7). Various other economic regionalization systems have been tried to improve management, like the *sovnarkhozy* of 1957 (varying in number from 95 to 105) or the 47 industrial management regions somewhat later. Recent emphasis has been on 'territorial production complexes', using spatial industrial linkages as growth poles. Such frequent changes suggest gross uncertainty over how to manage and stimulate an increasingly sluggish economy.

United States

Territorial administration in the United States dates from the late eighteenth century and expanded as new *states* joined the Union through westward colonizations. Apart from the older colonial lands in the east, many territorial divisions at all levels were fixed within arbitrary geometrical boundaries, often laid out well before the land was properly settled. The *states* enjoy considerable autonomy and their internal administration is defined in their constitutions, allowing variety, though the pattern is remarkably consistent considering the large size of the country and the diversity of its social, economic, and physical nature.

The usual immediate level below the *state* is the county, with some exceptions: Alaska is not divided overall in a county system; Rhode Island (no larger than many western counties) is not organized on a county basis for administration; parishes replace counties in Louisiana; and since 1960 there has been no active county administration in Connecticut. There are just over 3,000 districts of county type, but their area varies appreciably (the largest exceeds 51,700 km^2 in California and the smallest less than 250 km^2 in Virginia), with

Figure 6.7 USSR — *Makrorayony*

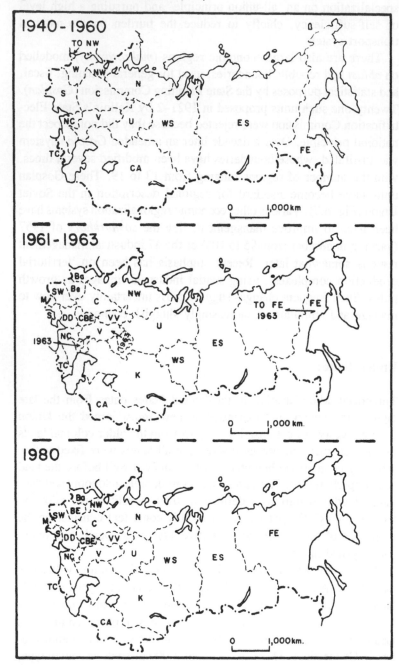

The changing boundaries of these major planning regions reflect efforts to adjust to developments in the spatial pattern of the economy

usually a crude relation to land value and population density. The number of counties in each *state* ranges between 3 in Delaware and 254 in Texas. Despite the great changes in the population and economy, county boundaries have been only modestly altered. Urban counties provide a far wider range of services than the predominantly rural ones.

Twenty-one *states*, wholly or in part, are divided into townships, covering New England and the Mid-West as well as the mid-Atlantic coast. As settlement spread, many original townships were divided, with irregular shapes reflecting the needs of early settlement and seldom exceeding 75 km^2. In the Mid-West, where cadastral survey preceded settlement, townships have characteristically rectangular shapes ('six miles by six miles'), even though many diverge from the regular 'congressional township'. Townships were here divided into square-mile sections for one to four farms, depending on terrain. Where plantations or estates, with a semi-feudal slave economy, predominated, or in the sparsely settled west, there was no place for the self-governing township. Townships are of diminishing importance, since they are generally too small to give the economies of larger units and the sense of community they represent weakens in an ever more mobile population (Fig. 6.8).

An urban community may be incorporated as a municipality by the *state* legislature, with its functions and limits defined by charter. As it grows, it may seek boundary extensions, though these often produce awkward and inefficient arrangements, complicating relations with adjoining authorities, as in Los Angeles. Municipalities presently number over 18,000 and range in population from a few hundred to several millions (e.g. New York) and in area from a few hectares to over 2,000 km^2 — the largest municipal areas are almost as great as the smallest *states*.

Because the individual *states* are responsible for the minor civil divisions, local government functions and patterns vary considerably. In some respects the United States reflects an internal territorial administration 'adapted' to local conditions. In such a free economy and society 'state' control at whatever level is possibly less significant than in the more regulated societies and economies of Europe, both east and west. Perhaps the most striking contrast with Europe is that American society, so committed to 'progress', is much more conservative in changing or reforming local government which in so many instances appears to Europeans to be ill-fitted to contemporary challenges and spatial patterns.

Figure 6.8 USA — states and townships

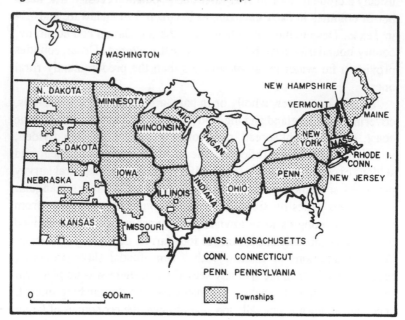

The townships are generally typical of the older states, well-reflected in this map

Some examples from the Third World

Territorial-administrative systems in Third World countries mostly reflect European influences, particularly in former colonial territories. There is, however, usually a marked 'traditional' element represented in the lowest echelons of local government. As most of these countries have societies and economies with a strong agrarian bias, the territorial-administrative structure shows a marked focus on rural organizations.

In Latin America, apart from federal structures like Brazil and Mexico, the influence of the former colonial administrative system can still be felt. Some countries (e.g. Bolivia) have made few changes in their territorial-administrative divisions since first becoming independent early last century, but others have made substantial adjustments (e.g. Chile under the Military Junta since 1973 has decentralized its administration and Peru carried out a major revision of its local-government system in 1984).

The newly independent countries in Africa have usually adapted their local-government structures from colonial days. Consequently former French territories still reveal a pattern of territorial administration imprinted by French practices and terminology. This is seen in Madagascar where the upper echelon bears a clear French imprint, but there is a traditional base, with the 'fokotany' communes, some 11,000 in number, managing village affairs. Above these, 92 sub-prefectures are grouped into 18 prefectures, themselves comprising 6 provinces. In former French Guinea the highest level of administration (some 33 *régions*) comprises four major divisions based on the principal geographical and ethnic elements of the country, viz. Guinée-Maritime, Moyen Guinée (the Futa Djallon Uplands), Haute Guinée, and Guinée Forestière. Former British colonial territories are usually marked by a particularly strong lower level of local government firmly in the hands of tribal chiefs or village headmen. This has been taken over from the British practice of seeking to preserve local institutions with a structure of indirect rule, first introduced by Lord Lugard in West Africa. Sierra Leone, for example, has 3 regions and one special area divided among 148 chiefdoms organized as local government units. In Malawi the colonial inheritance has remained fairly clear, with 3 regions and 24 districts, each administered by a district commissioner. Since Independence some countries have had remarkably volatile territorial-administrative systems, notably in the extremely complex ethnic pattern of Nigeria, where no consensus has been established on how to divide its large territory into federal units.

India since Independence has been a successful 'Union of States', with 22 *states* and 9 union territories. Below this upper echelon the Republic is divided into 40 municipal corporations, 1,274 municipalities, 815 town and 'notified' areas, and 62 'cantonment boards', each with wide-ranging local responsibilities. With three-quarters of the population living in the countryside, an elaborate three-tier rural administrative system is being spread through the *states*, allowing some adaptation in each to special local conditions. The system operates at village, block, and district level, each with members elected by and from villagers, looking after agricultural affairs, health, and some educational interests, while local justice is administered through village courts.

The State of Malaysia comprises former British holdings in the Malay Peninsula and in Borneo. It comprises three principal elements — Peninsular Malaysia, Sabah, and Sarawak — united by a strong central federal government and divided into thirteen *states*, each with its own constitution. The federal authority is supreme in all instances.

Nine of the *states* are under hereditary rulers, the others under governors, but the whole governmental system otherwise bears an imprint of British parliamentary traditions. The *states* are divided into districts, but in Sabah there is an intermediary level of 'residencies' and in Sarawak of 'divisions'; though both are becoming less important, they have so far preserved many features of the British colonial period. In Peninsular Malaysia the *mukim* is a rural unit forming a significant lower tier. Each *mukim* is usually a large area comprising one or several villages (*kampongs*), whose head (sometimes locally elected, sometimes appointed by the *state* government) is the main liaison between the villagers and the district administrator. Local government development was much encouraged in the early 1950s, especially in urban communities. The bodies that emerged varied considerably in degree of self-government and in designation. Five *states* (Johor, Kedah, Kelantan, Malacca, and Negri Sembilan) have no local bodies below district level and the erosion and abolition of the lower echelons have been explained as necessary to eradicate corruption and to improve administrative efficiency. It is possible this trend will extend to other *states*. The general tenor since Independence has been for the system in all its aspects to favour Malays over other peoples, in particular the Chinese.

Clearly, from the examples given, territorial-administrative systems represent a complex aspect of territorial management within states, designed variously to satisfy ideological criteria, economic management, population, settlement policy, or the manipulation of public opinion, while in the most liberal states they are encouraged to serve the public expression of local wishes.

7

The rise and demise of empires

Throughout history expansive and aggressive societies have sought to dominate other lands and peoples for prestige and power, generating the political structure known as 'empire'. Imperialistic urges may be underlain by various motives, usually in some combination, but the supposed political or strategic advantages are common ground, just as are the economic advantages believed to arise from empire. Imperial ventures may arise from a search for 'living space', often at the expense of the indigenous people of the conquered lands. Once embarked upon, imperial expansion has usually gained a momentum of its own, continuing until it overreaches its own resources, but, once momentum ceases, decline sets in quite soon.

Imperial control may be exercised indirectly through military threat, political pressure, or even offer of protection or subsidy, forcing recognition of one's overlordship on client states, vassals, or protectorates. Dependency often arises not just from direct conquest and subjugation but through connivance and acquiescence with the imperial power on the part of the subject territories in return for protection — how else could vast empires be built using limited human resources? Empire-building succeeds best where the skilful use of the psychological dimension is effectively understood. Imperial control may impose its own representatives on government and management at all levels in the 'colonial' territories or it may preserve much of the indigenous system and rule indirectly through a small elite. 'Colonization' may simply be an elite of officials and soldiers of the imperial administration or a partial settlement by colonists seizing the best land or establishing towns, but it may be a broad colonization, displacing, integrating, or effectively exterminating indigenous population. Empires may comprise compact and contiguous territories (as seen in the Russian and Chinese empires) or be scattered territories, like the overseas empires of European powers dependent on seapower. Considerable political geographical contrasts exist between the thalassocratic empires and compact territorial ones.

Early empires were often ephemeral personal ventures, seldom surviving long after their creator. Classic examples are the vast imperium built in his short life by Alexander the Great; the thalassocratic empire of Canute of Denmark; or the powerful Frankish Empire of Charles the Great. Personal empires were commonly overlordships, in which loyalty was paid by vassals to the 'emperor', whose leadership and personal charisma were as important as his military and political strength. More lasting empires were founded by family leadership, where the title of emperor was passed from generation to generation, and the dynasty became the focus of loyalty, so well illustrated by the polyglot empire built by the Habsburg in Central Europe.

THE MODEL OF EMPIRE — THE ROMANS

The Roman Empire grew less by any grand design than through victories over adversaries that left it holding additional territory, so acquisition was more the consequence than the purpose of frequent wars. Growth was thus rather piecemeal as the search for security was pursued: once one opponent had been subdued, another emerged, and Roman successes drew in more territory. The process begun in the Italian peninsula had spread to eliminate threats on the peripheries, in Gaul, in Rhaetia, and in Noricum, as well as the momentum carrying them into Africa, the Balkans, and Asia Minor, let alone Britannia (Fig. 7.1). Though Roman power came to encompass the shores of the Mediterranean, making it a Roman lake, the empire remained in essence continental.

As one success followed another, Roman problems of manpower and management multiplied. Nevertheless, so long as a dynamic, expanding empire was maintained, usually solutions to the immediate difficulties were found. Manpower became the most intractable difficulty, posing not only military problems but also in manning the civil administration. Much of the empire was governed indirectly by a small Roman staff (but there was nothing comparable to a colonial civil service) that kept existing native administration functioning and was particularly commissioned to raise taxes and recruit troops. It worked best in Greece and Asia Minor, where fairly advanced societies used to effective self-government existed, but

Figure 7.1 The Roman Empire at its zenith

Roman Empire

Temporary holdings or weak Roman control

0   500km

The strength and weakness of the Empire is reflected in its annular form around the Mediterranean — *Mare Nostrum*. Adapted from *Putzger Historischer Weltatlas*, Berlin 1979

was less successful in Gaul, in Spain, and in Germania, where tribal organization was backward, so much closer supervision was demanded.

The Roman Empire was hegemonic, with numerous client states and vassals, held to loyalty by subsidy or by punitive expeditions against those who stepped out of line. The disposition of the legions was as much for internal order as for frontier defence. Squabbles among the subject peoples were capitalized on by the Romans, playing one off against the other through diplomacy in the concept of 'divide et impera'. Because the Romans realized the dominant dimension of power was as much psychological as physical, they managed to control

161

large areas with relatively modest forces, depending on fear of the legions, whose mobility was assured by good roads and communications. They gambled on trouble always being limited to a few spots in the Empire rather than any general insurrection, for they had inadequate resources to maintain a proper strategic reserve. By tolerance and fairness, but a fierce and ruthless response to any troublemakers, they maintained a level of peace not previously experienced. This was a powerful force drawing the Empire together through the stability achieved, encouraging trade and reasonable living standards. The material attractions of Roman civilization generated willingness for romanization, exercised through the army and towns, with the possibility of the coveted award of Roman citizenship.

THE HOLY ROMAN EMPIRE

The Pope made a vain bid in AD 800, to keep the Empire (Fig. 7.2) alive by making the most powerful leader of the time the Holy Roman Emperor, charged to reconstitute the secular identity of the *Imperium Romanum* to match the spiritual power of the Church of Rome, which for a time achieved its goal of being a universalist church. The subsequent struggle for supremacy between the Pope and Emperor immensely weakened the concept, though the Empire remained a vague political force for a thousand years. Effectively the elected Emperor became a title for a German prince, but despite attempts it never became hereditary to one dynasty.

No Emperor managed to assert his effective central authority, forced to accept that kings and lesser rulers in reality governed themselves as they saw fit. The struggle between Pope and Emperor for supremacy for three centuries made Italian affairs the prime imperial concern, allowing German princes to go their own way, with excessive feudal fragmentation into petty states. In the quest for the phantom of Empire, the Emperors failed miserably to create a German monarchy and sense of nationhood, even though 'of the German nation' was added to the imperial title. By the mid-thirteenth century when they ceased meddling with any effect in Italian affairs, it was too late for the Emperors to unite Germany and an attempt to breathe life into Empire in the Golden Bull of 1356 merely 'legalised anarchy and called it a constitution' (Bryce 1889).

Figure 7.2 The Holy Roman Empire

Though its boundaries fluctuated, the Empire slowly contracted on the west and south until its dissolution in 1806. Adapted from several sources

THE DYNASTIC EMPIRE

German medieval colonization eastwards and southeastwards provided a cradle for two major empire-building dynasties — the Prussian Hohenzollern and Austrian Habsburg — each pursuing its endeavour within the vague framework of the Holy Roman Empire (Fig. 7.3).

The South German Habsburg began their rise to dynastic prominence when Count Rudolf was elected Holy Roman Emperor in 1273 and broke the power of the Bohemian Przemyslid dynasty. From 1438 until 1806 an almost unbroken run of Habsburg Holy Roman Emperors fortified the dynasty in its power struggle. Their real success lay in astute marriage policy, obstinate diplomacy, successful force of arms, and having the Church on their side, changing and adapting

Figure 7.3 The Habsburg and Hohenzollern Empires

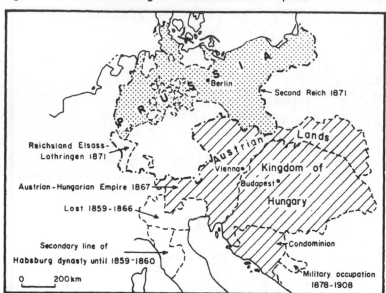

Both empires held commanding nodal positions. The Hohenzollern Empire (the Second *Reich*) was shortlived (1871–1918), while the Habsburg never achieved the ambition of their motto *Austriae est imperare orbi universo*

to political expediency so long as it furthered dynastic interest. Demanding absolute loyalty, the dynasty seldom turned to its subjects for popular support.

The widely spread Habsburg territories, bound neither by geography nor nationality, enjoyed no common name other than 'lands of the House of Habsburg'. Menaced by Napoleon, Franz II, King of Austria and of Hungary, in 1804 invented the appellation 'Emperor of Austria', an essentially dynastic title indicating the Empire of the House of Austria and *not* the Empire of the Austrians. In 1867, weakened and humiliated by Prussia, the Habsburg accepted the Magyar claim to partnership (building on a status long enjoyed) as the Empire of Austria-Hungary. In many respects the Habsburg empire was more a collection of vast entailed estates rather than a true state — all sentiment and endeavour focused upon the dynasty that represented only itself. The *Pragmatic Sanction* of 1713 defined the indivisibility of the Habsburg domains and a bureaucracy was created to execute its will. Expressing no national attachment, this essentially 'imperial' bureaucracy was managed in German by officials impressed by German ideas, culture, and traditions, identifying completely with a German dynasty, yet administering a polyglot empire, 45 per cent of whose subjects were Slav (1910), 23 per cent German, 19 per cent Magyar, and several small nationalities comprising the remaining 11 per cent. Among the dynasty's most loyal subjects were the urban Jews, whose Yiddish speech counted them as Germans in official statistics.

Three elements dominated the economy and society of the late Empire. Close to the dynasty, the aristocracy was socially and economically conservative, a powerful representative of agrarian interest. The latter were equally supported by the socially conservative but politically more radical peasants, greatly strengthened by reforms in the 1780s, the imprint of which outlasted the Empire to survive until the end of the Second World War. Townspeople, a progressive bourgeoisie of German and Jewish artisans and entrepreneurs, were the third element. Intellectuals stirred up nationalistic pressure for autonomy for the different peoples, creating a policy dilemma for the dynasty. Increasingly socially divided and economically weaker, whether it conceded or not spelled equal trouble for an Empire unlikely to stand the strain. The stable peasantry and powerful aristocracy struggling to maintain agrarian dominance had meant little encouragement for industrialization, made worse by modest endowment with mineral resources, so by 1900 the Empire could not match great-power status to economic fact. Such a trend made membership of the successful German *Zollverein* impossible, destroying the possibility that was proffered for leadership of *Deutschtum*. The First World War simply hastened the inevitable.

In the fifteenth century the South German Hohenzollern family inherited the Mark Brandenburg, a precarious political hold, surrounded by powerful, jealous neighbours. Though the Hohenzollern could consolidate their land territorially and politically, there was little they could do to compensate for its poor and unattractive quality; not until the seventeenth century could the dynasty make valuable gains, augmented further after the Thirty Years' War. In 1701 the ruler crowned himself King of Prussia and set about building the scattered and unpromising territories into a powerful state where the cardinal virtues were considered to be orderliness, simplicity, and thrift, while 'Prussia's armies were her frontiers'. Having contributed to Napoleon's downfall, Prussia emerged territorially enhanced by rich Rhenish lands and politically powerful enough to dominate North Germany and would have been even stronger, had there not been Habsburg opposition. Unfortunately its western and eastern territories were separated by a belt of petty states which it had every desire to eradicate.

The German Confederation created in 1815 under Austrian presidency proved impotent. Prussia, so much more truly German than the Habsburg Empire, was seen increasingly by German princes as their champion, especially as it strengthened its hold as a bastion for the defence of the German realm against France in the Rhineland. Prussia's defeat of Austria in 1866 confirmed its leadership, with substantial territorial gains and hegemony over a new North German Confederation. Great power status now went hand in hand with rising economic strength, demonstrated in the Prussian victory over France in 1871 when, as a result, at the request of the German princes (except expectedly Austria), the King of Prussia was declared Emperor of the Second *Reich*. Though concessions, more cosmetic than real, were made to South German particularism, power lay truly in a Prussian Empire.

The Romanov dynasty demanded absolute loyalty of all their varied subjects to the 'Tsar of all the Russias' (Fig. 7.4). Though the westwards expansion of Muscovy was largely hegemonic, the push into southern Russia and the Ukraine was a colonization, just as it was in the Volga basin and the Ural, but especially in Siberia, where colonization was not always by volunteers. The vastness of Siberia fell to the tsars because no other power could compete for such an inaccessible land. The spread into Caucasia was partly hegemonic, partly colonization, while in Central Asia many groups became vassals of the tsars, though Russian grip was consolidated by Slav colonization or where necessary by conquest, with a powerful arm of Russian imperial ambitions in the twelve Cossack 'armies' settled as soldier colonists in volatile areas. To no mean extent the Russian Empire was made by Slav colonization and native people, faced by integration or extinction (largely through disease or vodka!), believed 'the Russian comes as your friend and acts as your enemy'.

THE OTTOMAN EMPIRE

The Ottoman Turks built their empire in the fourteenth and fifteenth centuries on their military prowess, conquering or reducing to vassals much of the Middle East, North Africa, and the Arabian Peninsula, while they also spread into the Balkans and Danubia. Their expansion seemed unstoppable after the capture of Constantinople in 1453, until a coalition of Christian powers under the Polish king resoundingly defeated them before Vienna in 1683.

Figure 7.4 Growth of the Russian Empire.

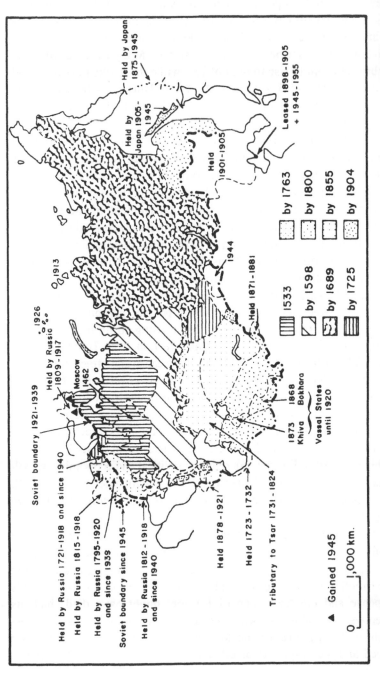

The map emphasizes the vast expansion virtually unopposed into Siberia and Central Asia. Adapted from *Atlas Istorii SSSR*, vols I–III, Moscow 1948–1954

The Turkish administration used Phanariot Greeks as agents, but they became a detested group, using their favoured status for oppression and usury of the masses as they, with increasing opportunism, grew richer. In Europe the broad masses were largely Christian, over whom the Muslim ruling elite exercised control through a highly trained and ruthless corps of janissaries (recruited mostly from Christian boys taken as tribute from their parents). Many Christian officials embraced Islam to rise to high office, while the Turks took over and adapted numerous Byzantine institutions. During its heyday in the fifteenth and sixteenth centuries, many Christian observers admired the Empire's efficiency.

The pastoral tradition of the Turks led them perhaps to view their subjects as a shepherd views his wealth in sheep, while as nomads they had little attachment to territory and consequently little interest in its development. Without skill or interest in economic matters, they were content to leave management of the economy to favoured subordinates but, as the Empire's economic problems grew from the sixteenth century, through stagnation, a rising population, and inflation, relations between the Turkish authorities and the Christian subjects in Europe deteriorated. Throughout their domains the early liberalism gave way to increasing despotism and merciless exploitation of the common people, perhaps a reflection of territorial expansion having placed a rising strain on administration, especially on effective central direction from Istanbul. Great financial and administrative tensions were cause and effect in a vicious round of neglect of simple actions of everyday government. Disregard for law and order allowed brigandage to flourish, dislocating trade and commerce, even though these were encouraged, despite being hampered by a lack of interest in establishing effective communications. As everyday standards and the general level of attainment fell behind the rest of Europe, not only the political and economic strength was sapped, but also the Turkish military machine deteriorated: with the expansive momentum of the Empire lost, more progressive opponents benefited.

Nineteenth-century independence movements, especially in the Balkans, took advantage of these embarrassments, encouraged by Turkey's enemies, the European great powers. Rebellion and insurrection helped the nationalists gain territorially at the Sultan's expense and the 'sick man of Europe' might have died, had his enemies been able to formulate an agreed policy for dismemberment of the Empire among themselves and their clients. Fired by greed and jealousy, one or other commonly stepped in to prevent

Figure 7.5 Spread of European Imperialism I.

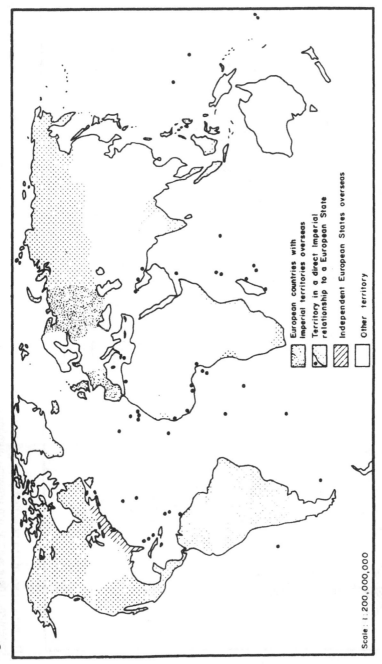

Scale : 1 : 200,000,000

The eighteenth century saw European countries already extending their influence overseas, with the American colonies establishing their independence after 1783

European countries with
Imperial territories overseas

Territory in a direct Imperial
relationship to a European State

Independent European States overseas

Other territory

Figure 7.6 Spread of European imperialism II.

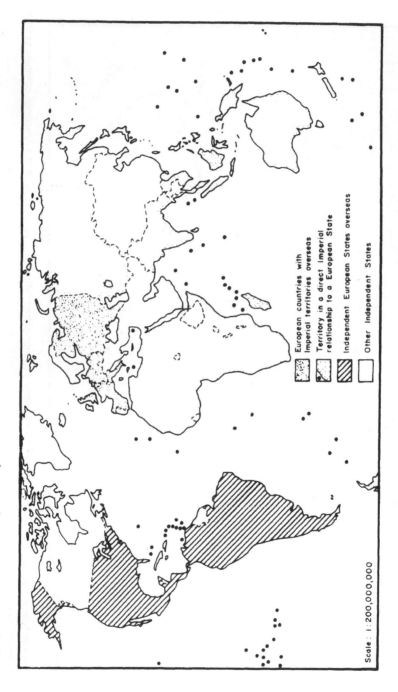

Scale : 1 : 200,000,000

European countries with
Imperial territories overseas

Territory in a direct Imperial
relationship to a European State

Independent European States overseas

Other Independent States

The early years of the twentieth century saw the maximum spread of European overseas empires and the growth of Russia's continental imperium. There was also a growing number of independent states founded by one-time colonial Europeans overseas a tasty Turkish morsel falling to a rival.

EUROPEAN EMPIRES OVERSEAS

The Renaissance generated an innovative and broader vision and a remarkable expansive and aggressive energy to explore the world. Growing stability and wealth encouraged trade and a search for new sources of raw materials and markets: interest focused particularly on increasing trade with the east, but the traditional trade routes by land were blocked by hostile powers, so new seaways lured explorers, aided by improved navigation and ships. New lands discovered by the voyagers were claimed and conquered for their sponsors.

Overseas empires of European countries grew rapidly in the sixteenth and seventeenth centuries (Figs. 7.5 and 7.6). Early leaders were Spain and Portugal, major Roman Catholic powers whose imperial ambitions were approved by the Pope in his division of the world between them in the Treaties of Tordesillas (1494) and Zaragoza (1529) for exploration and conquest. The imperial ventures of both powers were distinguished by a mixture of greed and piety. Spain was most anxious to obtain bullion from its overseas territories to replenish coffers emptied by expensive wars in Europe: Portugal was more interested in trade on a near-monopoly basis, particularly with the 'Spice Islands'. Whereas the Spanish penetrated deeply into Central and South America in a search for bullion, the Portuguese did little more than establish coastal trading bases in South America, Africa, India, and East Asia.

In Spanish America concentrations of Indians in favoured settlement areas became centres of extensive colonial territories, with new ones set up as Spanish control spread. Vice-royalties and captaincies-general multiplied, with an image of home developed in their focal areas by administrators and military men who became the new landowners. Because of the difficulties of communication, a remarkable degree of freedom to run their own affairs was enjoyed, so emissaries and orders from Spain were increasingly resented. Ibero-American society began dreaming of independence, successfully feeding early nineteenth-century revolts. Surviving Spanish possessions in the Caribbean and in the Far East fell to the United States in a brief war in 1898, though holdings in Africa survived until the late twentieth century.

Less ambitious, the Portuguese consolidated their hold on the

Brazilian coast, the seat of the Court during the Napoleonic Wars. In a compromise agreement Brazilian independence was accepted in 1825. In the late nineteenth-century scramble for Africa, Portugal's small coastal holdings were transformed into large territories as Angola and Mozambique.

The English and French soon followed, attracted particularly to North America, mostly ignored by the Iberians, where their conflict of interest formed much of their colonial history until the late eighteenth century. They also struggled over territories in the Caribbean and the Far East, where they clashed with Spain and Portugal.

The British were in the end the most successful empire-builders: at their imperial climax after 1918 their territories covered almost a quarter of the earth's land surface. Early interest in overseas possessions was stimulated by opportunities for trade in the close mercantilist system of the time, but there was also a desire for territories suitable for settlement, generated by Tudor fears of possible overpopulation at home, while mercantilism made tropical or sub-tropical plantation agriculture, using native or imported labour, attractive to investors. In the early stage raiding Spanish commerce in the Caribbean, using the islands as bases for piracy, was quietly encouraged by the home government, which unlike the other powers, preferred to leave acquisition and colonization to individuals or chartered companies. Even the subjugation of India was begun by a chartered company of remarkable diversity but surprisingly modest means. The British made little conscious effort to develop an 'empire', preferring to gain possessions piecemeal through the action of individual citizens, but even then accepting new territories only reluctantly. As British seapower grew, however, interest developed in holding strategically located offshore or oceanic islands.

Outside the tropics Britons proved successful settlers, even though many went to escape constraints of home society, for adventure, or hope of personal fortune. This was first epitomized in the North American colonies where a remarkable social and political cohesion developed. Through the home government's inflexibility, lack of interest, and aloofness, the colonists ultimately broke away in the American War of Independence. Their victory brought tacit recognition that long-term overseas possessions should be led towards independent management of their own affairs (Durham Report 1832) in the expression 'good government is no substitute for self-government'. British colonists in Canada developed their own thrusting community alongside the more conservative and *dirigiste* French colonization on the St Lawrence. Equally successfully the British settled in Australia and New Zealand, though the initial use of Australia as a penal colony reflected sadly on the home government's attitude to such territories. A less happy relationship arose between British settlers and earlier Boer colonists in South Africa. This century these territories have all become independent nations in their own right.

Despite the loss of the American colonies, the empire continued to grow but, with the industrial revolution, substitution of free trade for mercantilism made overseas trade penetration a greater priority than merely imperial conquest. Like the Romans the British faced a manpower problem in running their empire — garrisons were small, relying on seapower to bring reinforcements in emergencies, and British settlers were encouraged to organize their own defence, although the rise of British seapower made any challenge from other powers harder. Administration was modest, often adapting native institutions (where these were reasonably developed), though practices distasteful to Europeans were eradicated. This was notably institutionalized in India, where the British presence begun by the East India Company was strengthened by making contracts with over three hundred princes as 'protectorates', though they largely continued their own government. India's manpower and military traditions were moulded into a splendid colonial army as the sub-continent became a key strategic base, turning the Indian Ocean into a 'British lake'. In Africa Lugard introduced 'indirect rule' in Nigeria, to be followed elsewhere. Whereas French colonial policy sought to turn selected Africans into Frenchmen, Lugard was convinced European influences should refashion, enrich, but not overwhelm African traditions. Much of the British empire (notably in Africa) had a hegemonic character, comprising 'protectorates' where native rulers accepted guarantees of British protection against outside interference in exchange for British management of their foreign policy and defence and guidance in other matters. Internal administration remained in native hands, but the territory was opened preferentially to British commerce.

Colonial territories changed hands several times in frequent wars, especially in the West Indies, with rivalry between the British, French, and Dutch. There was some attraction in fighting one's battles in distant waters, particularly for the stronger naval power. The Dutch began empire-building late in the sixteenth century, the zenith of their commercial and maritime vigour. Interest focused on the trade potential of the East Indian Spice Islands, from which they drove the

Portuguese and barred the British. They secured their seaways by several small holdings, most importantly the Cape of Good Hope and by control of Ceylon. They also spread into the Caribbean. Their regime was exploitative, run by elite administrators ruling indirectly in Indonesia through native institutions, with planters holding much authority. The Dutch were autocratic and centralized (despite considerable welfare and economic achievements in later times) and never envisaged any self-government for their territories. Their imperial ambitions were curbed by the decline of their maritime power from the late seventeenth century.

The heavy continental commitment in Europe meant France could never devote itself as wholeheartedly as Britain to seapower. Well organized and with a *dirigiste* government, France established a successful presence on the St Lawrence in North America, spreading its influence into the Mississippi Basin and Great Lakes. For a time this threatened to contain the English colonies and take over the vast interior. Somehow the French never became as thrusting and anxious to settle their new lands as the British and their North American initiative petered out. Napoleon's defeat stripped France of most overseas possessions, retaining only a few West Indian islands, but later in the century a new, larger empire was built, beginning with the occupation of Algeria (in an effort to stamp out North African piracy). Defeat in the war of 1870 gave a new stimulus, with territories gained in Africa, Indo-China, and Madagascar. In Morocco France demonstrated skilful diplomacy in maintaining some semblance of the country's traditions and usages.

The motives for acquiring overseas possessions were seen by the French more clearly and analytically, though perhaps less honestly, than by the British. Colonial resources were needed to redress the French position in Europe as German economic upsurge and political unity occurred after 1870. The military potential of colonial manpower was particularly attractive and imperial ventures gave a new self-confidence. All French colonial subjects enjoyed the opportunity to rise to high political or administrative positions, provided they adopted French culture and civilization. In this policy of 'assimilation' colonial peoples with the will and ability to merge into French civilization were accepted as partners, but apart from plural societies in some island possessions it did not work well. Subject peoples with their own strong cultures (like Islamic communities in North Africa) were resentful. Seeing failure, the French after 1918 shifted to a concept of 'association', in which native institutions would be used but without conscious effort to preserve them, while only a small elite would be encouraged to adopt French language and culture, and all would be managed by a vigorous and highly centralized control from Paris.

Nationalism in Germany and Italy encouraged imperial dreams, believing overseas colonies brought economic fortune but also an image of power and prestige. Relatively little territory remained unclaimed to satisfy these ambitions, generated by popular clamour, so there was a rush to grab parts of Africa. German interest was primarily in territories as suppliers of raw materials for industrialization, but the Italians, greedy to reconstitute the image of Rome, wanted colonies for settlers to ease supposed overpopulation at home. Russia quietly intensified its hold on the interior of Asia, even though it had sold Alaska to the United States in 1867, but this expansion created grave misgivings for Britain and threatened a delicate balance of interest in China.

The late nineteenth-century British had little alternative but to compete for the remaining territories, especially as French aspirations had sinister strategic undertones and German naval ambitions were disquieting. Moreover, the lessons of history showed that an empire, once established, took on an expansive impetus of its own, as each acquisition provided strategic and other reasons to absorb adjacent territories. An awakening in Britain itself of popular imperialism imparted a new pride and interest in empire, now becoming a conscious goal as never before. The new imperial powers wanted absolute sovereignty, monopoly, and exclusion in their colonies, so the British felt obliged to oppose such neo-mercantilism by defending their own concept of free trade by direct competition. Nevertheless, more so than their rivals, the British built a new indirect imperialism through heavy investment in independent countries, such as railway development in Latin America. Even so, British imperial policy was diffident and hesitant: had it exerted itself fully, it could have probably seized more of Africa and excluded competitors from the Pacific and Southeast Asia.

In the early stages many small countries sought overseas possessions but were usually quickly overtaken by larger predators. In the seventeenth century Brandenburg-Prussia briefly held trading posts on the West African coast and a century later there was an ephemeral Swedish presence in the West Indies. Denmark built a boreal empire in Iceland, Greenland, and Faroe, but also held the West Indian Virgin Islands as well. Following Stanley's explorations an international consortium, the Congo Association, had been formed, becoming the Congo Free State in 1885, virtually the personal property of the Belgian king. Its ruthless exploitation caused an international scandal, so, when the king died, the Belgian government reluctantly assumed control, seeking to live down the previous poor image.

EMPIRES IN DECLINE

European imperialism never regained its momentum after 1918. Britain, France, and Japan carved up German and Turkish possessions, though the gains mostly were 'mandates', territories managed on behalf of the League of Nations until ready for independence. In the Middle East Jewish and Arab nationalist rivalry involved the mandate holders in serious civil disorder. Italy, however, disgruntled because it had made few gains, embarked on a new imperialism with an unprovoked attack on Abyssinia. Britain between the world wars pursued the long-expressed aim of increasing autonomy to parts of the Empire ready for it. The process was slow because autonomy or eventual independence could only be conceded when there was a sound basis of democratic parliamentary government. The granting of sovereignty to the so-called 'White Dominions' (Canada, Australia, New Zealand, and South Africa) under the 1931 Statute of Westminster recast the Empire as the British Commonwealth, though Egypt and Iraq edged out, as did the Irish Free State.

After the Second World War the European overseas empires were quickly dismembered. In Southeast Asia the interval between the Japanese collapse and the return of the old imperial authorities was long enough for local people to feel independence. This was a key factor in the French inability to regain a firm hold over Indo-China and forced Holland to concede independence to Indonesia, while it also speeded independence from Britain of Malaya, Singapore, Borneo, and Burma. The haste with which the Indian sub-continent gained freedom led to misgivings that too little time had been allowed to solve the age-old differences between Hindus and Muslims, confirmed in Partition between India and a new state concept of Pakistan. Incompatibility between the two parts of Pakistan led to separation of the eastern, Bengali lands as the new republic of Bangladesh in 1974.

Events in Asia intensified pressure for independence elsewhere, notably in Africa. In British West Africa indirect rule had left government in native hands at all but the highest level, for which a native elite was being developed. Despite misgivings that local people were insufficiently trained for top government, independence was steadily conceded from the mid-1950s, but subsequent instability in Nigeria and

176

Ghana perhaps confirmed the fears. The path in East Africa was less straightforward, for in Kenya and Southern Rhodesia there were considerable European settlements. Even so, in the early 1960s, independence was achieved by Kenya, Tanganyika (Tanzania), Uganda, and Nyasaland (Malawi) along with Northern Rhodesia (Zambia), after failure of a brief and abortive Central African Federation. Former protectorates were also freed.

Conflicting philosophies of empire affected developments in Southern Rhodesia and in South Africa. The former was the creation of the maverick diamond-magnate Rhodes as the chartered British South Africa Company (until made a Crown Colony in 1923 with limited self-government). It was settled by independently minded European colonists convinced by Rhodes' sentiments on private initiative and tinted by the Boer philosophy of the white man's superiority for ever over the natives. To stave off sharing power in a constitutional arrangement acceptable to London, as applied elsewhere, the white Rhodesians declared their unilateral independence in 1965 (with South African connivance). Not until 1979 could a constitutional formula be found, giving the black majority power and yet apparently safeguarding the interests of the European colonists in the new Republic of Zimbabwe. In South Africa, despite their subjugation by the British, the Boers were never reconciled to the relative liberalism of the long-term aims of British policy. Despite South Africa receiving Dominion status in 1931, the growth of Boer influence saw racial segregation and indefinite white supremacy and domination accepted as tenets of policy. Refusal to bow to pressure for a more racially egalitarian approach was the catalyst for the declaration of a republic outside the Commonwealth in 1961.

The growing number of territories gaining independence changed the Commonwealth as constituted in the 1930s from an imperial into an international organization, playing quite a different role in world affairs (discussed in Chapter 8).

Although the overseas territories remained loyal after the German occupation of metropolitan France in 1940, pressures for eventual independence began to mount. In West Africa relations between France and the newly independent states created in the 1960s have been generally cordial, but French rule in Algeria was ousted in a violent civil war, where reluctance to concede independence had been conditioned by the presence of many French colonists. Communist insurgency took over in Indo-China.

The Second World War stripped Italy of its empire, confirmed in transference of power to native administrations after 1945, though retaining a trusteeship of Somalia until independence in 1960. Independence sometimes came, however, through political difficulties arising at home, as in the easy-going Portuguese administration in Angola and Mozambique that survived until 1975, when pressures for independence could no longer be resisted by a weakened home government. The promotion of no other policy than to better the home country's resources from these territories left them unprepared for independent management and they have fallen prey to civil war. In Angola Cuban troops have sought to re-establish government control, just as Belgian forces had to intervene in the civil disorder after independence in the Congo. The British have also had to retain a garrison in Belize (ex-British Honduras) to protect it from its predatory neighbour, Guatemala.

PROBLEMS OF INDEPENDENT EX-COLONIAL STATES

Establishing a stable society and flourishing economy in countries emerging to independence from imperial systems has as much depended on cultural and historical factors as on the territory's natural endowment. Since independence in the late eighteenth century the thirteen American colonies, the core of the United States, have grown into the world's most powerful and vigorous nation-state. In contrast, only moderate success attended the collection of nation-states that emerged in the early nineteenth century as the Spanish and Portuguese empires in the Americas collapsed. This century Canada, Australia, and New Zealand slowly loosened their imperial ties to Britain to become successful nation-states in their own right. Economically successful but with violent internal social tensions of its own making, has been South Africa.

Successful post-imperial nation-states have not been limited to communities descended from a major influx of European colonists. India, for example, with its own powerful civilization before falling under British control, successfully adapted to independence after 1947: we may also cite Egypt, Iraq, Syria, or Sri Lanka, and Pakistan (since shedding its incompatible eastern part, now Bangladesh). Industrious and gifted Chinese in a plural society have contributed substantially to making Malaysia and Singapore among the most successful parts of Southeast Asia. Within the French sphere comparable developments have come in Morocco, Algeria, Tunisia, and Madagascar.

In Africa 'decolonization' has brought problems when applied hastily among territories hardly ready for the responsibilities of independence. Political independence has often been constrained by economic factors, since economic ties, for historical and commercial reasons have usually remained close to the former imperial power, with notable dependence for financial, technical, and other aid. Unfortunately aid has not always been wisely used, often being spent on unnecessarily large armed forces, on dubious 'prestige' projects (a state airline, for example), or squandered through mismanagement or corruption. Aid-donating powers, in competition with one another for ulterior political motives, have been reluctant to monitor or control the use of their aid, particularly as they have often had little real interest in the true welfare of the recipient. Some aid and technical assistance have been inappropriate, perhaps often through donors failing to appreciate the real needs: African and Asian countries at a low level of development might profit more from Chinese rather than from Russian or American experience! Some new African countries have pursued unsuitable policies in their anxiety to earn foreign currency to buy luxury imports for the ruling elite, while, even where development has achieved rising living standards, these have frequently been eroded by a failure to contain explosive population growth.

Rapid 'decolonization' in Africa and Asia has commonly been mirrored subsequently in political instability. One-party pseudo-democratic regimes or outright totalitarianism have been rife (Malawi, Zimbabwe, Uganda), while several countries have been torn by internal strife, with long, bitter campaigns slowly destroying the inherited economic infrastructure (Sudan, Zaire, Mozambique). Such commotion has often had a powerful ideological bias, fuelled by covert superpower rivalries (Angola), with conscious 'destabilization' sought by superpowers in some strategically important areas (the Horn of Africa). The instability and volatility of so many newly independent countries have seriously deterred outside investors.

Ex-colonial boundaries have often lumped together peoples with little affinity to form new independent nation-states. Ethnic diversity, for example, in Mauretania means that the Moorish majority has to live with a third of their countrymen more akin to Negro Senegal. Asians in Uganda, introduced by the British for railway and other government work, had become a significant commercial and professional element, vital to the country's future, but were expelled after independence in a programme of 'africanization'. The Organisation for African Unity's decision not to support pressures and claims for territorial adjustments, while aimed at encourging peace and stability, may not long term be the ideal approach, leaving increasingly frustrated 'irredenta'. Sadly, we must also recognize that some internal stresses are an inheritance from a divide-and-rule policy of imperial days.

Imperialism remains alive but less overt. Consider the hegemonic imperialism of the Soviet Union in Eastern Europe, where the nation-states retain all the trappings of sovereignty but in reality are manipulated through a massive Soviet proconsular system. In the capitalist world, despite vague 'national' labels, immense transnational industrial corporations encroach on national sovereignty. Seeking to optimize their trading activities by switching operations from one country to another to maximize locational advantages, the scale of their investment and operational decisions influences even quite large national economies. Consequently government policy changes have to consider the likely reaction of such 'multinationals'. Major inter-national financial institutions operating in the volatile investment and currency markets also exert influence on national policy through their expression of 'confidence' or otherwise, although these activities can be monitored and to some degree regulated by national central banks acting in concert. International 'confidence' expressed through invest-ment decisions or currency parities has come to have the impact once exerted by political or military pressures on internal policies, as seen in international action to constrain Third World indebtedness by insistence on specific economic or other pre-conditions before debts can be rescheduled.

It is perhaps unreal, unwise, and unfair to expect the newly independent ex-colonial nation-states to mirror European concepts of organization and expression of the nation-state. These new countries have their own values, attitudes, traditions, and expectations, though we should expect them to honour by priority the needs of their peoples in economic and social well-being and to accept the usually practised standards of international behaviour.

8

Aspects of international organization

Since countries do not and cannot exist in isolation, they organize their relations with others for mutual benefit, expressed through formal treaty organizations for particular objectives encompassing regional, continental, or even global groupings. International organizations have formal constitutions imposing responsibilities, duties, and even financial obligations on members, while some may have legal entitlement to enact their own laws, with penalties or expulsion for non-observance. Any organization is, of course, only as successful as its participants make it and may fall apart through dissent or sheer inertia. Success and longevity usually depend upon the strength of common purpose among members.

International organizations have three main objectives: political, to strengthen the influence of a group in the corridors of world power; economic or social, to better the welfare of its members; or military, to provide mutual defence against enemies. Though an organization may express one of these as its primary objective, one or both of the others may also be influential in its activities. Small countries commonly form organizations for their mutual benefit, but usually seek to draw in a major power for added strength, while large powers may sponsor such organizations to attract smaller powers into their influence. International organizations may be contained in one or other of the superpower *blocs* (i.e. the capitalist world led by the United States or the socialist *bloc* dominated by the Soviet Union) or they may be spread across both *blocs*, but some have come to represent the emergent grouping of 'non-aligned' powers.

INTERNATIONAL ORGANIZATION EXEMPLIFIED BY EUROPE

Europe well illustrates the problems of creating an effective and durable international organization. At the Congress of Vienna in 1815 the powers allied against Napoleon sought to prevent domination of the Continent by any one power through the concept of the Concert of Europe, the idea of acting together to settle disputes peacefully. This was embodied in the Quadruple Alliance of Britain, Prussia, Russia, and Austria, initially for twenty years, with periodic meetings for 'consideration of the measures most salutary for the maintenance of peace'. The enigmatic Tsar Alexander also drew Prussia and Austria into a 'Holy Alliance' with Russia, a 'Christian union of charity, peace and love', eventually, it was hoped, to be 'one Christian nation'. Ultimately most European rulers joined (except England's Prince Regent, the Pope and, not surprisingly, the Turkish Sultan), though few took it seriously.

As the century progressed, the fragile peace was maintained by an elaborate, shifting system of alliances, mostly military in character, that sought to preserve the *status quo* through the balance of power, manipulated by a diplomatic chess game, blocking unwelcome moves by rivals. It became most complex after mid-century, preserving an armed peace where the implied threat of war could be averted by the system remaining flexible enough to adjust readily to shifting rivalries. After 1871 the structure centred around French reluctance to accept the new strength of Germany which undermined their leading role in Europe dating back to the sixteenth century. Russia and Austria were sympathetic to Germany as conservative powers anxious to maintain the existing social and state systems against liberal trends emanating notably from France, and the new power balance was accepted by Britain and by Italy. No diplomatic or military partner could be found for France as a counterweight, for France was unpopular through clashes with others (notably Britain) over imperial interests overseas. Russia's Balkan intentions alarmed Austria as did those in the Ottoman Empire and Asia alarm Britain. By the 1890s Germany, Britain, and Italy formed an entente with Austria-Hungary against Russia, which was supported by France, and Europe moved towards the new century in three *blocs* — the Triple Alliance (Germany, Austria-Hungry, and Italy), the British Empire (somewhat by itself), and the Franco-Russian alliance. This albeit rather unstable structure was, however, flexible enough to allow endless diplomatic manoeuvre to preserve peace.

Early this century, Germany, by now the most powerful

Continental power, embarked on policies that triggered fears that it sought (and could achieve) outright domination in Europe. German naval policy was an unacceptable challenge to Britain which by 1907, having solved outstanding imperial frictions, sided with France and Russia. The threefold structure had become a less flexible bipolar one — Britain, France, and Russia on the one hand and Germany and Austria-Hungary on the other. The alignment was seen in London, Paris, and St Petersburg as 'containment' of German policy: in Berlin, as dangerous 'encirclement'. The situation became more volatile when the Balkan Wars of 1912–13 replaced a decadent but stabilizing Ottoman Empire by mutually antagonistic petty states. By 1914 Austria-Hungary had decided on a once-and-for-all solution of the Balkan Question that precipitated the end of this system of alliances by the Great War.

Britain, France, the United States, and their wartime allies wanted after the Great War to continue co-operation and draw in other countries to help maintain the peace following 'the war to end wars'. The concept and legal basis of the so-called *League of Nations* were quickly put together in 1919 by Wilson, the American President, with high ideals but little corresponding experience or realism. Membership was dominantly but not exclusively European. The start was inauspicious as opposition in the Senate to Wilson meant that America failed to join and Germany, a key country, was excluded until 1927. The League was seriously weakened because it had no means to enforce its decisions (frequently ignored or flouted) and because it made no effective headway in arms control: it failed miserably in the Manchurian Incident of 1931 and the Italian attack on Abyssinia in 1936. Few states had much confidence in the League and chose to act mostly outside its framework. The British wre unenthusiastic and the Germans lukewarm, feeling that it was dominated by France, and after Germany's withdrawal in 1933 it steadily lost credibility and was to all intents moribund by the late 1930s. It did, however, manage the solution of a few imperial problems through the idea of 'mandated territories' (cynics saw these simply as disguised annexations); it also created some successful international agencies concerned with health and labour and helped to better the lot of minorities.

There was little confidence in the League and a new system of military treaties proliferated, mostly between the Succession States of East Central Europe against possible German or Hungarian revisionism or the fear of 'Bolshevik Russia', and usually backed by France in the hope of winning back leadership. In 1925, in the Locarno Treaty, Britain guaranteed the security of the Franco-German and

183

Belgo-German frontiers and the demilitarized Rhineland against either French or German aggression, while France did the same for Czechoslovak and Polish frontiers with Germany, which was admitted to the League. Like earlier counterparts, these treaties, while promising armed resistance, represented little more than consultation between general staffs. As Germany grew stronger after 1933, this French-inspired security system disintegrated.

A strong desire to try to create a powerful regional agency with a vigorous policy of promoting political and economic harmony swept Europe again after 1945. The keynote was political commitment, even some readiness to surrender sovereignty, and the hopes were embodied in the European Movement which in 1947 at the Hague founded the parliamentary Council of Europe with its seat in Strasbourg. Progress foundered on diverging views of how quickly matters should advance as old European attitudes surfaced again: France wanted rapid advance, perhaps seeing an opportunity to seize political leadership, whereas the British, always wary of Continental enthusiasms, preferred slower progress, relying on traditional channels of action. They also feared commitment might jeopardize the destiny they felt still lay in Commonwealth leadership.

In the deadlock the French, suspected of self-interest, sensed a loss of support and changed their approach, made necessary anyway as the *Wirtschaftswunder* clearly destined West Germany to renewed strength. French uneasiness about German economic potential made a search for rapprochement rather than traditional hostile rivalry an unavoidable option, while the Germans, internationally disgraced, were most willing to grasp anything that might help re-establish their place and image in Europe. The successful customs union (Benelux) since 1948 between Belgium, the Netherlands, and Luxemburg demonstrated a way for close co-operation and a proposal to integrate heavy industry to speed reconstruction in France, Germany, and Benelux was pursued. Italy joined as well, but Britain (seen as a potential honest broker in Franco-German reconciliation) declined. The European Coal and Steel Community (ECSC) set up in 1951 was envisaged as the first of several specialized authorities with 'limited functions but real powers' representing a breakthrough to supranational organization. Having its own legal powers and sanctions, it was a counter to the 'economic nationalism' that bedevilled interwar Europe.

Primarily an expression of collective defence against possible Soviet intentions, but also covering social and economic co-operation, Britain, France, and Benelux had signed the Brussels Treaty in 1948.

This had been instrumental in encouraging America to form a broader Western defence system, incorporating itself, Canada, the Brussels Treaty members, and other European countries in the North Atlantic Treaty Organization (NATO) (1949). This was a practical expression of the Truman Doctrine pledging American help to all free peoples fighting aggression and oppression (implicitly Soviet), but also a distinct shift from the long-standing American isolationism, making Europe a forward garrison and glacis for 'Fortress America'. Unlike earlier military treaties, NATO has seen close integration of members' forces in equipment and training and in a multinational command structure.

The success of NATO and the ECSC encouraged the Brussels Treaty powers to establish a European Defence Community (EDC), with integrated multinational armed forces as a forerunner for a corresponding political integration. Its creation was made a prerequisite to granting full sovereignty to the German Federal Republic whose new armed forces would be integrated within it. This made the French fearful of the unknown quantity of new German forces and they failed to ratify the EDC treaty (1954). To avoid the serious political repercussions of failure, a stopgap intergovernmental Western European Union was proposed, in which Britain would enhance its military presence and the original Brussels Treaty would be modified to allow accession of West Germany, whose forces were admitted to NATO (1955). German forces were important because any forward defence strategy involved West German territory, though sadly NATO was later weakened by withdrawal of the French at the zenith of de Gaulle's chauvinism (1966).

As political integration was proving so difficult, the members of the successful ECSC decided to proceed further towards economic integration. Britain again decided not to join them, preferring a 'pragmatic' rather than their 'constitutional' solution, and favouring a loose association in a free-trade area instead of the six members' idea of a formal customs union with closely codified co-operation. Basically it was a difference between 'consensus' and 'supranationalism'. Nevertheless, Britain's continued imperial decline and mounting economic embarrassment made closer ties desirable to an increasingly economically successful Continent, but without jeopardizing leadership of an ever more independently minded Commonwealth. Other partners were found who felt formal supranationalism too binding or who were otherwise unable to join such an arrangements and in 1960 Britain, Norway, Sweden, Denmark, Austria, Switzerland, and Portugal established the European Free Trade Association (EFTA), a much looser organization, to which later Iceland joined and Finland became an associate. It worked well in freeing trade in industrial products and in reducing trade barriers, notably with countries within the Organization for European Economic Co-operation (OEEC) (founded in 1947 to administer the American's European Recovery Programme), whose own success had seen it expanded in 1961 to the Organization for Economic Co-operation and Development (OECD) on a global basis.

Somewhat more quickly (1956) France, Germany, Italy, and Benelux in the Treaty of Rome instituted the European Economic Community (EEC) ('Common Market') cast in the image of its time and of its sponsors. The six founding countries were a relatively close-knit, interdependent group, with a common historical experience and a common aspiration: the economic well-being and security of Benelux were tied closely to Franco-German harmony; France saw a 'common market' as a means of rapprochement, hopefully defusing any political tensions with Germany, which itself viewed the treaty as a significant route to returning to its proper place in Europe; for Italy it made Cinderella more hopeful of finding a prince! For cynics, the Treaty was just a trade-off between French farmers and German industrialists.

The achievements of the Treaty of Rome have been uneven. Industrial integration, a relatively successful sphere, has doubtless been aided by the nature of industrial firms, with their wide, multinational contacts and organization. It has regulated fair trading and set technical and quality standards, though some 'harmonization' has seemed petty, designed for no better reason than unnecessary uniformity. That the Treaty would solve the ills of European farming proved illusory, for the Common Agricultural Policy has been too inflexible and the measures really needed have often been too radical to be acceptable in a politically most sensitive sector, but the whole gamut of difficulties has been compounded by the sheer diversity of farming itself. Embarrassing surpluses bought under guaranteed prices have brought the Community to the brink of insolvency. Large industrial countries importing food, like Germany and Britain (since accession in 1973), make massive financial payments to support farmers in France and Italy where reform still lags, and now in Greece and Iberia (accession in 1986). The high stability price on the Continent (where protectionism always created high food prices) has been a serious burden in Britain, where until entry a cheap food policy had been the norm, and the raising of Community tariff barriers to food imports has generated political tensions with the Commonwealth, notably Australia and New Zealand. It was overcoming those same

barriers that provided the attraction, however, of Community membership for Denmark and Ireland in 1973 (Fig. 8.1).

Though a European Currency Unit has been accepted for accounting purposes, and since 1979 some members have sought to limit
variations in parities between their national currencies in the European
Monetary System ('the Snake'), political reasons as much as any have

Figure 8.1 International organization in Europe.

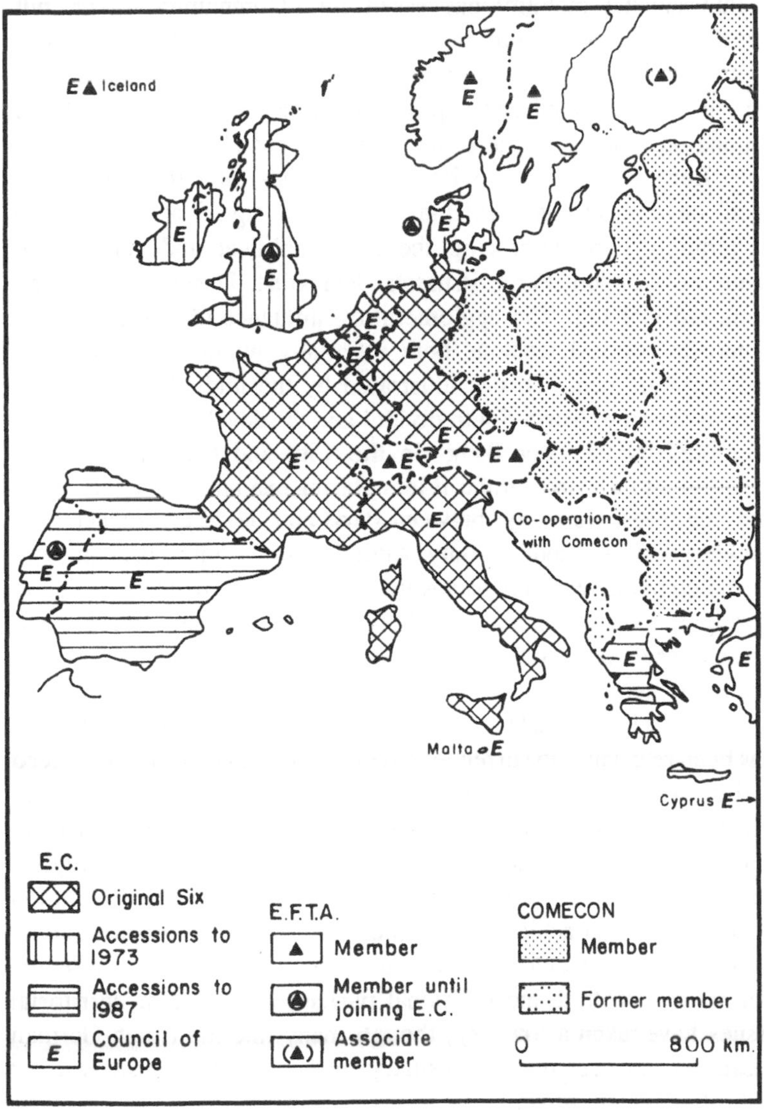

E.C.
Original Six
Accessions to 1973
Accessions to 1987
E Council of Europe

E.F.T.A.
▲ Member
◉ Member until joining E.C.
(▲) Associate member

COMECON
Member
Former member

0   800 km.

Adapted from various sources

187

preserved national currencies. Dismantling exchange controls has, however, helped financial institutions in member countries and the greater measure of accord between their central banks, under some political duress, has also been a benefit. Certainly Community funding of regional aid has helped less favoured areas and ameliorated the core–periphery contrasts, though an imminent political storm has blown up (1987) with proposals to channel regional aid principally to southern European members.

International mobility has eased intra-Community travel, but has done little for labour mobility. It has, however, had considerable repercussions on locational patterns as internal tariff barriers have been dismantled and adjustment made from relatively closed national economies to the broad transnational Community pattern. The 'Eurodelta' (Rhine, Schelde, and Maas) and the Rhine axis itself have taken on particular economic significance within a Community core embracing the great political and commercial centres of London, Paris, Brussels, Frankfurt, Düsseldorf, and the Randstad Holland. The agglomerating influence of this Euro-core generates political tensions both internally within member states and between them. This is illustrated by the mid-1980s debate in Britain over the 'North-South divide' and the development problems generated by the 'Big Bang' in the City and the 'M25 Golden Ring' — we may find similar if less glaring examples between North and Southwest Germany, between Paris and 'le désert français', or between north and south in Italy. Added tensions will be brought with accession of the relatively backward economies of Greece, Spain, and Portugal. Full 'harmonization' between all members in 1992 will present challenges sufficient to recast radically their social and economic geography.

The principal obstacle to integration, whether political or economic, has been reluctance to surrender sovereignty even marginally. Indeed anti-marketeers in Britain have made sovereignty the central issue. The right to veto proposals on grounds of 'national interest' has not been sufficient to avoid claims of infringement of sovereignty on several occasions. Though joint consultation and agreed common action have been achieved in numerous matters, common political or foreign policy would demand a surrender of sovereignty to a degree still unacceptable to most national parliaments. At least, territorial issues have taken a low key, though many like the Dutch-German disagreement over the Ems estuary or the Anglo-Spanish issue of Gibraltar remain alive.

188

COMECON

The *Council for Mutual Economic Assistance* (CMEA or Comecon) is the socialist *bloc's* very different counterpart to the European Economic Community. Having forced Poland, Czechoslavakia, and Hungary not to participate in the European Recovery Programme, Stalin had to offer some compensation, if not to impress his satellites, certainly as a gesture to the world at large. The Council's existence was unexpectedly announced in 1949 with membership open to any country willing to accept Soviet-style economic management. All the East European socialist *bloc* countries joined, though Albania (under Chinese influence) subsequently defaulted. Later Mongolia, Cuba, and Vietnam were to join, while Yugoslavia has been an observer. The aim was to strengthen economic ties by mutual assistance between members and in particular with the Soviet Union as creator and dominant partner, but there was no supranational dimension, with emphasis on self-containment and self-sufficiency of each national economy. The Council was advisory and consultative: through it information and expertise would be exchanged and trade promoted, essentially bilaterally, mostly with the Soviet Union. In reality relations continued as immediately post-war, focused on or radiating from Moscow, with Soviet pre-eminence maintained by manipulation by the massive Soviet proconsular system.

During the 1960s the Council began to take form, though still dominated by Moscow, as it was moulded increasingly by pressure for modernization and sophistication of their economies from Eastern Europe. This had to be done in a framework acceptable to Moscow and some went too quickly, with dire consequences as in Czechoslovakia in 1968, but the Soviet Union became receptive to financial and trade mechanisms more akin to capitalist practice. National autarky was jettisoned for the 'international socialist division of labour', whereby members specialized in producing what they could manufacture best and then traded between themselves to exchange surpluses to cover deficiencies. Stalinist autarky had been an unreal objective for members other than the Soviet Union, since none had a domestic market large enough to maximize economies of scale nor anything like an adequate natural resource endowment, with short-comings made good by the Soviet Union, whose economic hold thus strengthened its political hegemony. This was replaced by encourage-ment of joint projects between members but usually with a Soviet input, while most members have been cajoled into participation in development projects in Siberia and the Ural.

189

Accession of West Germany to NATO was made the occasion by the Soviet Union to create a defence organization, the Warsaw Pact, broader than existing bilateral treaties with the East European client states. This made little difference to the military situation, for the satellite armed forces were already integrated into the overall Soviet defence machine.

It would be surprising if considerable changes did not occur in Comecon in the wake of Gorbachev's *glasnost* and *perestroyka* policies in the Soviet Union. It is yet (1988) too early to see what form such change will take.

THE UNITED NATIONS ORGANIZATION

The Second World War reawakened the desire for an effective forum in which to defuse international tensions. Such a body was initiated by the United States, the United Kingdom, and China, with the Soviet Union (which was also allowed to have Byelorussia and the Ukraine as voting members to strengthen its position) at San Francisco in 1945, when fifty countries signed the Charter of the *United Nations Organization* (UNO). Open to all peace-loving countries, membership is now virtually universal, while so far no member has left or been expelled. It is a more powerful organization than the defunct League of Nations, provided with more formidable teeth and much greater resources, so it can undertake tasks impossible for its predecessor, compared to whom it has a much wider brief. The most important organ has been the Security Council which has monitored international disputes and arbitrated in their solution, though the five permanent members (United States, United Kingdom, France, Soviet Union, China) hold a veto. Unlike the League, UNO has been capable of enforcing its decisions, because it can call on peace-keeping forces under its own management, the troops being on secondment from member countries. In this respect the multinational United Nations armed forces in Korea, 1951–3, did much to establish its credibility. Though the Security Council, the General Assembly, and other sections seem slow and ponderous, they have proved valuable 'talking shops' where in lengthy debate international crises can in most instances be defused. The UNO has undoubtedly helped in smoothing the 'decolonization' process in Africa and Asia by assisting newly independent countries to find economic and political stability through a wide range of specialized agencies. The growth in the number of independent Third World countries has made them a powerful lobby

190

in the Organization against the more established industrial nations of America and Eurasia.

THE COMMONWEALTH

The product of the unusual view of imperial development accepted in Britain after the defection of the American colonies in the 1780s (Chapter 7), the concept of a British Commonwealth of Nations arose at the Imperial Conference of 1927. It crystallized in the Statute of Westminster in 1931, defined as a group of 'autonomous communities within the British Empire, equal in status, in no way subordinate to each other in any respect of their domestic or foreign affairs though united by common allegiance to the Crown and freely associated' (Fig, 8.2).

The modern Commonwealth emerged in 1949 when members accepted India's decision to become a republic and yet retain full membership. At that time the king was accepted as the symbol of free association of its member nations and as such as head of the Commonwealth. As more territories have become independent, it has formed a strange congerie of nations with tenuous common links: former status as British imperial possessions; English as a *lingua franca*; legal systems based on British usage; and, it is hoped, British-style parliamentary democracy. Most important is the conviction that membership confers worthwhile benefits, though some of the former elements have been lost, notably trade preferences and free immigration into Britain, while monies under the Development and Welfare Fund are no longer so freely available. As nations consolidate their identity and memories of imperial traditions and days fade, it is difficult to see what long-term future awaits it, with a clearly widening gulf between Britain and the old 'White Dominions' on the one hand and on the other the newly independent countries in Africa and Asia which more and more do not live up the to the democratic standards Britain would prefer. As a political lobby the Commonwealth is being weakened as divisions of opinion between members appear to multiply. It is difficult to believe that Britain's destiny any longer lies with the Commonwealth, a view persisting into the 1960s that probably cost Britain the more profitable leadership of Europe. In 1985 eighteen Commonwealth members remained the 'Queen's realms', twenty-six were republics, and five were indigenous monarchies. Not all former British territories have retained membership on independence.

191

Figure 8.2 The Commonwealth

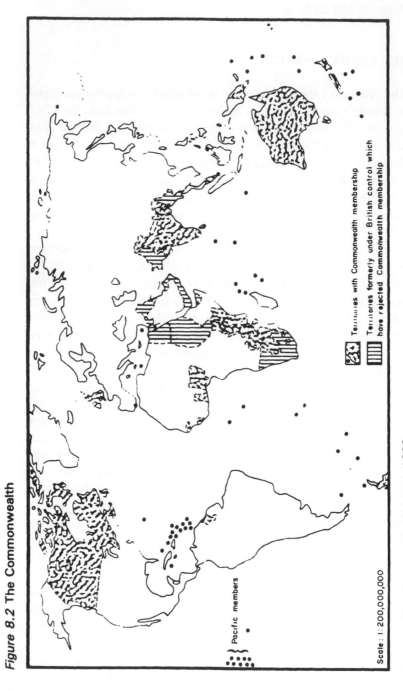

Scale : 1 : 200,000,000

Territories with Commonwealth membership

Territories formerly under British control which
have rejected Commonwealth membership

Pacific members

Based on The Statesman's Yearbook 1986

ARAB LEAGUE

The awakening of Arab-Islamic identity generated in some Arab leaders during the First World War hopes of a Greater Arabia arising from the ruins of the Ottoman Empire. However, the diversity of the Arab peoples and their regional interests failed to provide a firm enough common base and they were subsumed mostly into territories under British, French, or Italian control, beside a few notionally independent states. To create a greater sense of coherence, seven independent Arab states in 1945 formed the Arab League and membership ultimately grew to fifteen until Egypt was expelled in 1979 for its unilateral peace treaty with Israel (Fig. 8.3). Before the 1973 Israeli War the League did little other than mount a political and economic campaign against Israel and this continues, but it has subsequently begun to promote economic development among its poorer members by channelling to them some of the wealth earned by more fortunate members from oil, and it has offered assistance to Third World countries seriously affected by the oil crisis aftermath. The League has also sought to promote stability within its members'

Figure 8.3 The Arab League

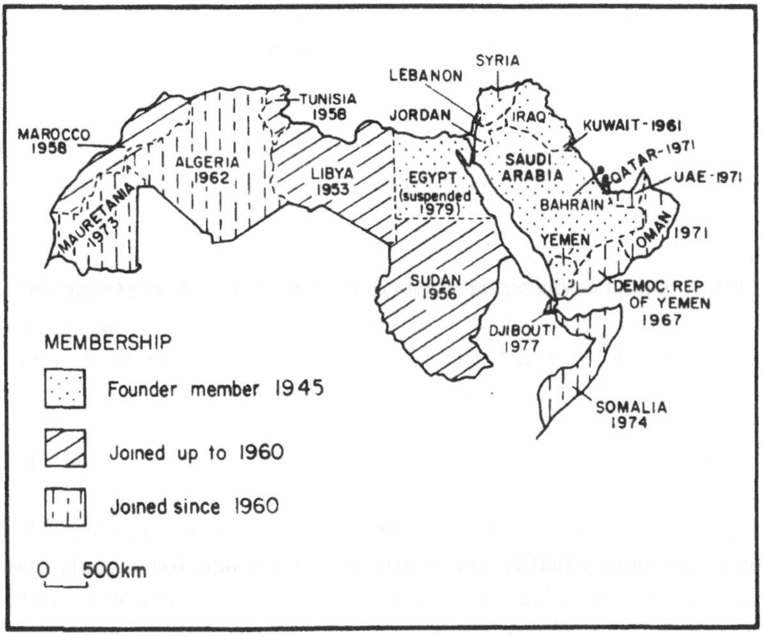

Based on *The Statesman's Yearbook 1986*

193

territories, notably through a predominantly Syrian force trying to keep warring factions apart in the Lebanon.

ORGANIZATION OF PETROLEUM EXPORTING COUNTRIES (OPEC)

In the 1980s eight of the thirteen members of OPEC are Arab-Islamic countries, even though any petroleum-exporting country may apply for membership. After 1945 nationalism in many countries brought pressure for ownership (if not management) of their oil-producing industries, which until then had been almost entirely controlled by American and European firms, although Mexico had nationalized its industry in 1938. The postwar catalyst was nationalization of the Iranian oil industry in 1951, followed in 1960 by the foundation of OPEC, with four of the five sponsors Middle Eastern countries. OPEC seeks to safeguard, individually and collectively, the interests of members in stabilizing oil prices and negotiating a fair return. Effort is directed in such a volatile commodity market to eliminating wide and frequent price fluctuations. The power of this small group was demonstrated in 1973 when it suddenly quadrupled crude oil prices and applied an embargo on Israel and its suppliers, gravely disrupting the world economy (even lapping over into the Soviet *bloc*). There is no doubt political motives generated in the Arab-Islamic world colour many OPEC moves.

ORGANIZATIONS IN THE AMERICAS

One of the older regional associations, with its origins dating back to 1890, is the *Organization of American States*. The modest objective of the original Pan-American Union to foster understanding among the nations of the western hemisphere has been subsequently considerably broadened. The present organization was created in 1948 to co-ordinate all formerly independent oganizations in the Inter-American system and to clarify and codify their relationships. The aim is to strengthen the solidarity of the Americas by drawing members together in defence of their territorial integrity and sovereignty. The basis is absolute equality among members, each state having only one vote in the Council. The Organization has directed much effort towards economic and social development in Latin America, where economic associations have developed widely, facilitated by the Economic

194

Commission for Latin America of the United Nations. These bodies have included the Latin American Free Trade Association (1961) and its successor, the Latin American Integration Association, while in 1960 the Central American Common Market was created and followed by the Andean states forming the *Grupo Andino* in 1969. Apart from a number of small local groupings, in 1975 twenty-five Latin American and Caribbean countries formed the *Sistema Económico Latinoamericano* for co-operation in economic and social matters at both intraregional and extraregional level. The Caribbean Community (Caricom) formed in 1973 aims at a 'common market' through economic co-operation; co-operation in such areas as health, education, technology, and culture; and co-ordination of foreign policy, but members do not necessarily participate in all three sectors.

ORGANIZATION OF AFRICAN UNITY

The *Organization of African Unity* (OAU) is an example of a regional grouping of states of mixed origin but common interest. Founded by thirty-two African countries in 1963, with the objective of furthering solidarity by co-ordinating political, economic, social, defence, and other affairs, it has acted as a significant lobby for African interests, particularly aiming to eradicate 'colonialism'. It has been reasonably successful in stabilization, assisting containment and solution of disputes among members. Importantly it enunciated a principle that territorial claims should not be made, even though boundaries of some countries are inappropriate delimitations inherited from the colonial past. It has successfully intervened where armed conflict threatened between members (Morocco-Algeria, Somalia-Ethiopia, Ghana-Upper Volta), but less successfully in civil war in the Congo, Nigeria, and the Sudan or in disputes involving forces from outside Africa, as in Angola and the Ogaden. Unfortunately it has been tarnished by the destabilizing influence of its Liberation Committees providing financial and other aid to insurgency. Though it professes to undertake activities of a non-political kind in promoting science and culture, these are rudimentary compared to its overwhelming political dimension.

INTERNATIONAL ORGANIZATION IN SOUTHEAST ASIA

The size and importance of Asia does not seem properly reflected

195

in its regional organization. Nevertheless, one of the most successful of all international efforts has been the Colombo Plan. Founded in 1950 by seven Commonwealth countries, it has grown to truly international proportions with twenty-six members. The aim has been to promote development of newly independent Asian countries. While overseeing development problems of Asia and the Pacific in a multilateral approach and co-ordinated manner, all negotiations for assistance are done bilaterally between the donor and recipient countries.

The Association of South East Asian Nations (ASEAN) was formed in 1967 by Thailand, Indonesia, Malaysia, Singapore, and the Philippines, while Brunei acceded in 1984. While aiming to speed economic growth and social and cultural development, it seeks especially to promote stability in Southeast Asia. It was slow to develop, having apparently done very little until the Communist takeover in Indo-China in 1975, but it has since been more vigorous. It has become a collaborative lobby for regional mutual interests and has acted to promote intra-regional trade and the region's trade with the outside world, while it has stimulated economic development projects seen as bulwarks against Communism. The Association has no competence to settle disputes either between members or with outside powers, but pressures exist from some members for it to develop a more vigorous common policy. Though drawn together by common fears and weaknesses, the Association's development has been hampered by nationalistic selfishness among the members.

Volatile political affiliations weakened the brief life of two military pacts in Asia, reflecting the common problem of international organization. Product of the Collective Defence Treaty of 1954, the South East Asia Treaty Organisation (SEATO) was a response to the collapse of the French in Indo-China, but it was soon undermined as Britain and France began to withdraw from imperial commitments in the region and Australia and New Zealand showed little enthusiasm. SEATO was ineffective during the Vietnam crisis when the region's security needed it most and began to dissolve soon after the war. The Central Treaty Organisation (CENTO) of 1955 had a similar fate, having begun as a mutual assistance agreement between Iraq and Turkey, subsequently joined by Iran and Pakistan, with active support from Britain and the United States. A serious blow was dealt when Iraq withdrew in 1959, but inactivity in Pakistan and a loss of interest by Iran and Turkey let the Treaty fall apart. It did, however, achieve some economic development within a strategic framework, notably completion of the Turkey–Iran rail link.

WORLD ECONOMIC ORGANIZATIONS

International organizations with principally economic objectives have grown in importance as the world's countries in the capitalist sector have sought to level out fluctuations in economic fortunes. After the financial traumas of the interwar period, effort was made in 1945 to manage the world monetary system through international agreement. Though primarily an economic function, exchange rates and monetary movements can have significant political dimensions. The two main organizations are the International Monetary Fund (IMF) and the International Bank for Reconstruction and Development ('World Bank'). Each IMF member is allocated a monetary quota both in relation to its contribution to the Fund and to its right to draw on it (these may be supplemented by borrowing). The purpose of the Fund is to promote international monetary co-operation and exchange-rate stability, with the object of expanding world trade by a multilateral payments system and freedom from exchange restrictions. It can assist members to surmount serious balance of payments difficulties, though stringent conditions are usually attached. Many poorer countries baled out by the Fund have been vociferous in claiming that such conditions have worsened their economic and social plight. The Board of Governors has delegated many powers to twenty-two executive directors, six appointed by the United States, Britain, West Germany, France, Japan, and Saudi Arabia (reflecting where a powerful financial muscle is held in the world today). The remaining sixteen are elected as representatives of groups of countries, whose votes are cast by their director.

The World Bank gives technical advice and investment funds for economic development among its poorer members. It has regional consultative groups and missions to ensure avoidance of wasteful overlap and help plan the most effective use of resources in fields such as agriculture, rural development, industry, and infrastructural projects. It works closely with regional banking institutions and with the banking system of COMECON.

GENERAL AGREEMENT ON TARIFFS AND TRADE

The Western world felt re-establishment of a free but fair trading system was a prerequisite to rebuilding the world economy after 1945. Negotiations began for creation of an international trade organization as a complementary partner to international monetary management.

197

At Geneva in 1947, a General Agreement on Tariffs and Trade (GATT) was signed and ratified, intended as an interim measure until an International Trade Organization could be instituted. This was done in 1948 at Havana, but only one of the twenty-three participants ratified its charter. Consequently GATT has remained the only international instrument defining trading principles for the overwhelming bulk of world trade, representing 90 countries, with a further 31 participating under special arrangements.

GATT is in effect a multilateral treaty defining a common code of practice for international trading as well as a forum where trade disputes may be settled by negotiation and consultation. It seeks to reduce trade barriers and is a modern version of the long-established 'most favoured nation' concept, with rounds of negotiations every five years or so to reduce tariffs further. The status of a member as a 'most favoured nation' must be negotiated and is commonly an occasion for political concessions as well.

GATT has unquestionably helped the expansion of world trade, mostly to the advantage of already rich industrial countries, while former colonial territories have seen less benefit. This has worsened the terms of trade for the latter as the price of industrial goods has risen more sharply than that of the commodities of which they are the primary producers. Third World countries felt little inclination to co-operate with GATT unless it were modified to surmount the widening gap between rich and poor. By the late 1960s the Third World countries had become sufficiently numerous to press the UNO for change. Out of the deliberations emerged the UN Conference on Trade and Development (UNCTAD), mainly seeking to use trading policy as an instrument for economic development.

Emergence of a wide range of international organizations, whether political or commercial, has been a marker in political geography this century and may well point a way ahead. Earlier chapters have charted the evolution and change in other principal elements of political geography and, as in all human institutions, change will no doubt continue. Apparently peaking earlier this century, nationalism as a driving force of group identity seems to be losing some of its force as modern technology shrinks distance and enhances exchange of information. Even though pressures for recognition as nations continue from many groups, what other bonds of group identity might possibly replace the nation? Certainly, too, as internationally generated lobbies and pressures mount, the nation-state, the most common form today of political territorial organization, has been increasingly forced to heed them. There is evidence to suggest states' freedom of action is being more and more constrained as one government's policies have to take greater cognizance of those of others, nowhere more clearly reflected than in economic and financial matters. Nevertheless, when it comes to joining international action and organization, the greatest obstacle is usually the reluctance of states to surrender, even marginally, any sovereignty. What will such change mean for the nature and strength of these institutions and especially for the territorial division of the earth's surface among them? This is particularly important as environmental and resource management becomes more pressing. It is hard to select the most likely scenario.

Selected bibliography

Adami, V. (1923) *National Frontiers in Relation to International Law*, Oxford: Oxford University Press.

Alexander, L.M. (1963) *Offshore Geography of Northwestern Europe*, Chicago: Rand McNally.

Ancel, J. (1936) *Géopolitique*, Paris: Delagrave.

—— (1938) *La Géographie des Frontières*, Paris: Delagrave.

—— (1940) *Manuel de la Géographie Politique*, Paris: Delagrave.

Aubin, H. (1929) 'Das deutsche Volk in seinen Stämmen — Überblick über die Besiedlung des Deutschen Bodens', *Volk und Reich der Deutschen*, Band I, Berlin: de Gruyter.

Barzini, L. (1984) *The Europeans*, London: Penguin.

Bateman, M. and Riley, R. (1986) *The Geography of Defence*, London: Croom Helm.

Beetham, D. (1984) 'The future of the nation-state' in McLennan, G., Held, D. and Hall, S. (eds) *The Idea of the Modern State*, Milton Keynes: Open University Press.

Bergesen, A. (ed.) (1980) *Studies of the Modern World System*, New York: Academic Press.

Bergman, E.F. (1975) *Modern Political Geography*, Dubuque, Iowa: Brown.

Bindoff, S.T. (1945) *The Scheldt Question*, London: Allen & Unwin.

Birch, A.H. (1977) *Political Integration and Disintegration in the British Isles*, London: Allen & Unwin.

Boateng, E.A. (1978) *A Political Geography of Africa*, Cambridge: Cambridge University Press.

Boggs, S.W. (1930) 'Delimitations of the territorial sea', *American Journal of International Law*, 24: 541–55.

—— (1937) 'Problems of water boundary definition', *Geographical Review*, 27: 445–56.

—— (1940) *International Boundaries — A Study of Boundary Functions and Problems*, New York: Columbia University Press.

—— (1951) 'National claims in adjacent Seas', *Geographical Review*, 41: 185–209.

Bowett, D.W. (1970) *The Law of International Institutions*, London: Stevens.

Bowman, I. (1928) *The New World*, 4th edn, Chicago: World Books.

—— (1942) 'Geography versus geopolitics', *Geographical Review*, 32: 646–58.

—— (1946) 'The Strategy of territorial decisions', *Foreign Affairs*, 24: 177–95.

Brand, J. (1974) *Local Government Reform in England 1888–1974*, London: Croom Helm.

Brierly, J.L. (1955) *The Law of Nations*, Oxford: Oxford University Press.

Brigham, A.P. (1919) 'Principles in the Determination of Boundaries', *Geographical Review*, 7: 201–19.

—— (1961) *British Foreign Policy*, London: Central Office of Information.

Brunn, S. (1984) 'The future of the nation-state system' in Taylor, P.J. and House, J. (eds) *Political Geography — Recent Advances and Future Directions*, London: Croom Helm.

Bryce, J. (1889) *The Holy Roman Empire*, London: Macmillan.

Buchanan, W. and Cantril, H. (1953) *How Nations See Each Other*, Urbana, Ill: University of Illinois Press.

Buckholts, P. (1966) *Political Geography*, New York: Ronald Press.

Burghardt, A.F. (1962) *Borderland — A Historical and Geographical Study of Burgenland, Austria*, Madison, Wis.: University of Wisconsin Press.

—— (1969) 'The core concept in political geography — a definition of terms', *Canadian Geographer*, 63: 349–53.

—— (1973) 'The bases of territorial claims', *Geographical Review*, 63: 225–45.

Burnett, A.D. and Taylor, P.J. (eds) (1981) *Political Studies from Spatial Perspectives*, Chichester: Wiley.

Burton, B. (ed.) (1967) *Problems of Smaller Territories*, London: Athlone Press.

Busteed, M.A. (1975) *Geography and Voting Behaviour*, Oxford: Oxford University Press.

Cahnmann, W. (1949) 'Frontiers between east and west in Europe', *Geographical Review*, 38: 615–25.

Carver, M. (1980) *War since 1945*, London: Weidenfeld & Nicolson.

—— (1983) *A Policy for Peace*, London: Faber.

Chadwick, H.M. (1945) *The Nationalities of Europe and the Growth of National Ideologies*, Cambridge: Cambridge University Press.

Chandler, D.G. (1974) *The Art of Warfare on Land*, London: Hamlyn.

Church, R.J.H. (1951) *Modern Colonialism*, London: Hutchinson.

Cipolla, C. (1973) *The Economic Decline of Empires*, London: Methuen.

Claessen, H.J.M. and Skalnik, P. (eds) (1978) *The Early State*, Den Haag: Mouton.

Clausewitz, C. von (1943) *Principles of War* (ed. Gatzke, H.W.), London: Bodley Head.

Clayton, A. (1987) *The British Empire as a Seapower*, London: Macmillan.

Cobban, A. (1969) *The Nation-State and National Self-determination*, London: Collins.

Cohen, S. (1973) *Geography and Politics in a World Divided*, 2nd edn, Oxford: Oxford University Press.

Colby, C.C. (1938) *Geographical Aspects of International Relations*, Chicago: University of Chicago Press.

Colombo, C.J. (1959) *The International Law of the Sea*, London: Longman.

Coplin, W.D. (1965) 'International law and assumptions about the state system', *World Politics*, 17, 615–34.

Cornish, V. (1936) *Borderlands of Language in Europe and their Relation to the Historic Frontier of Christendom*, London: Harrap.

Couper, A.D. (1978) *The Law of the Sea*, London: Macmillan.

Curzon, G.N. (1908) *Frontiers*, Oxford: Oxford University Press.

Darby, H.C. (1932) 'The mediaeval sea-state', *Scottish Geographical Magazine*, 47: 136–49.

Darwin, J. (1987) *Britain and Decolonization*, London: Macmillan.

Daveau, J. (1959) *Les Régions frontalières de la Montagne Jurassienne*, Paris: Presses Universitaires de France.

Davis, J.W. (1922) 'The unguarded boundary', *Geographical Review*, 12: 586–601.

201

Day, A.J. (1982) *Border and Territorial Disputes*, London: Longman.

Day, W. (1949) 'The relative permanence of former boundaries in Africa', *Scottish Geographical Magazine*, 65: 57–72.

Dayan, M. (1955) 'Israel's border and security problems', *Foreign Affairs*, 33: 250–67.

de Gaulle, C. (1934) *L'Armée de l'Avenir*, Paris: Colin.

Delbrück, H. (1900–1936) *Geschichte der Kriegskunst im Rahmen der politischen Geschichte*, 7 vols, Berlin: Stilke.

de Madariaga, S. (1952) *Portrait of Europe*, London: Hollis & Carter.

de Rougemont, D. (1965) *The Meaning of Europe*, London: Macmillan.

de Seversky, A.P. (1950) *Air Power — Key to Survival*, New York: Simon & Schuster.

Deutsch, K.W. (1952) 'The growth of nations — some recurrent patterns of political and social integration', *World Politics*, 5: 168–95.

Deutsch, K.W. and Foltz, W.J. (1963) *Nation-Building*, New York: Atherton Press.

Dewdney, J.C. (1967) 'Patterns and problems of regionalization in the USSR', *Research Papers*, 8: Department of Geography, University of Durham.

Dikshit, R.D. (1970) 'Toward a generic approach in political geography', *Tijdschrift voor Economische en Sociale Geografie*, 61: 242–5.

—— (1971) 'Geography and federalism', *Annals of Association of American Geographers*, 61: 97–115.

Dominian, L. (1917) *The Frontiers of Language and Nationality in Europe*, New York: American Geographical Society.

Earle, E.M. (1944) *Makers of Modern Strategy*, Princeton, NJ: Princeton University Press.

East, W.G. (1960) 'Political geography of land-locked states', *Transactions of Institute of British Geographers*, 28: 1–22.

—— (1961) 'The concept and political status of the shatter zone' in N.J.G. Pounds (ed.) *Geographical Essays on Eastern Europe*, Bloomington, Ind.: Indiana University Press.

East, W.G. and Moodie, A.E. (eds) (1957) *This Changing World — Studies in Political Geography*, London: Harrap.

East, W.G. and Prescott, J.R.V. (1975) *Our Fragmented World*, London: Macmillan.

Easton, S.C. (1964) *The Rise and Fall of Western Colonialism*, London: Praeger.

Eisenstadt, S.N. and Rokkan, S. (eds) (1973) *Building States and Nations*, Beverly Hills, Calif: Sage.

Elton, C. (1945) *Imperial Commonwealth*, London: Collins.

Fairgreave, J. (1915) *Geography and World Power*, London: University of London Press.

Falls, C. (1948) 'Geography and war strategy', *Geographical Journal*, 112: 4–17.

—— (1961) *The Art of War from Napoleon to the Present Day*, Oxford: Oxford University Press.

Farran, D. d'O. (1955) 'International enclaves and the question of state servitude', *International and Comparative Law Quarterly*, 4: 294–307.

Fawcett, C.B. (1918) *Frontiers — A Study in Political Geography*, Oxford: Oxford University Press.

Febvre, L. (1966) *A Geographical Introduction to History*, London: Routledge & Kegan Paul.
Fieldhouse, D.K. (1966) *The Colonial Empires — A Comparative Study from the Eighteenth Century*, London: Weidenfeld & Nicolson.
Fifield, R.H. and Pearcy, G.E. (1944) *Geopolitics in Principle and Practice*, Boston: Atheneum Press.
Fischer, E. (1949) 'On Boundaries', *World Politics*, 1: 196–222.
Fisher, C.A. (1950) 'The expansion of Japan — a study in oriental geopolitics', *Geographical Journal*, 115, 4–6 and 179–93.
—— (ed.) (1968) *Essays in Political Geography*, London: Methuen.
Fitzgerald, W. (1946) *The New Europe*, London: Methuen.
Förster, H. (1980) 'Zur Raumwirksamkeit der Integration in Osteuropa', *Fragenkreise*, 23546: Paderborn: Schöningh.
Franke, W. (1968) 'Die deutsch-dänische Grenze in ihrem Einfluss auf die Differenzierung der Kulturlandschaft', *Forschungen zur Deutschen Landeskunde*: 172.
Fraser, T.G. (1984) *Partition in Ireland, India and Palestine*, London: Macmillan.
Freeman, T.W. (1968) *Geography and Regional Administration*, London: Hutchinson.
Fuller, J.F.C. (1961) *The Conduct of War 1789–1961*, London: Eyre & Spottiswoods.
Geipel, J. (1969) *The Europeans — An Ethno-historical Survey*, London: Longman.
Gellner, E. (1983) *Nations and Nationalism*, Oxford: Blackwell.
Gilbert, E.W. (1939) 'Practical regionalism in England and Wales', *Geographical Journal*, 94: 29–44.
—— (1948) 'The boundaries of local government areas' *Geographical Journal*: 111, 172–206.
Gilfillan, S.C. (1924) 'European political boundaries', *Political Science Quarterly*, 39: 458–84.
Glassner, M.I. and de Blij, H. (1980) *Systematic Political Geography*, New York: Wiley (see also 1st edn and 2nd edn of 1967 and 1973 by de Blij alone).
Gottman, J. (1952) *La Politique des États et leur Géographie*, Paris: Hachette.
—— (1973) *The Significance of Territory*, Charlottsville, Va.: University of Virginia Press.
—— (1980) *Centre and Periphery — Spatial Variations in Politics*, Chichester: Wiley.
—— (1982) 'The basic problem of political geography — the organization of space and the search for stability', *Tijdschrift voor Economische en Sociale Geografie*, 73: 340–9.
Gray, C.S. (1977) *The Geopolitics of the Nuclear Era — Heartland, Rimland and the Technological Revolution*, New York: Crane Rossack.
Gregory, J.W. (1931) *Race as a Political Factor*, London: Watts.
Grimal, M. (1978) *Decolonization*, London: Routledge & Kegan Paul.
György, A. (1944) *Geopolitics — The New German Science*, Berkeley, Calif.: University of California Press.
Hall, H.D. (1948) 'Zones of the international frontier', *Geographical Review*, 38: 615–25.

203

Hall, W.E. (1924) *A Treatise on International Law*, Oxford: Oxford University Press.

Halpern, M.H. (1971) *Defense Strategies in the Seventies*, New York: Little Brown.

Harbeson, R.W. (1959) 'Transportation — Achilles Heel of National Security', *Political Science Quarterly*, 74: 1–20.

Hartshorne, R. (1933) 'Geographic and political boundaries in Upper Silesia', *Annals of Association of American Geographers*, 23: 145–228.

—— (1936) 'Suggestion on the terminology of political boundaries', *Mitteilungen des Vereins der Geographen an der Universität Leipzig*, Heft 14–15: 180–92.

—— (1937) 'The Polish corridor', *Journal of Geography*, 36: 161–76.

—— (1940) 'The Concept of "raison d'être" and maturity of states', *Annals of Association of American Geographers*, 30: 59–60.

—— (1950) 'The Franco-German boundary of 1871', *World Politics*, 2: 221–34.

Haupert, J.S. (1959) 'The impact of geographic location upon Sweden as a Baltic power', *Journal of Geography*, 58: 5–14.

Haushofer, A. (1951) *Allgemeine Politische Geographie und Geopolitik*, Heidelberg, Vowinckel.

Haushofer, K. (1939) *Grenzen in ihrer geographischen und politischen Bedeutung*, 2 Ausgabe, Berlin: Vowinckel.

Hay, D. (1957) 'Europe — The emergence of an idea', *Edinburgh University Publications in History*, 7.

Hay, R. (1955) 'The Persian Gulf states and their boundary problems', *Geographical Journal*, 120: 433–43.

Hechter, M. (1975) *Internal Colonization*, London: Routledge & Kegan Paul.

Hennig, R. (1931) *Geopolitik*, Leipzig: Teubner.

Héraud, G. (1968) *Peuples et Langues d'Europe*, Paris: Denoel.

Hertslet, E. (1875–1891) *The Map of Europe by Treaty*, 4 vols, London: HMSO.

Hertz, F. (1944) *Nationality in History and Politics*, London: Routledge & Kegan Paul.

Herz, J.H. (1957) 'The Rise and demise of the territorial state', *World Politics*, 9: 473–93.

Hill, N. (1945) *Claims to Territory in International Law and Relations*, Oxford: Oxford University Press.

Hobson, J.A. (1938) *Imperialism — A Study*, London: Allen & Unwin.

Hoffman, G.W. (1954) 'Boundary problems in Europe', *Annals of Association of American Geographers*, 44: 102–7.

Hofmann, H.H. (1969) 'Grenzen und Kernräume in Franken' in *Historische Raumforschung, 7 — Grenzbildende Faktoren in der Geschichte*, Hannover: Akademie für Raumforschung und Landesplanung.

Hoggart, R. and Johnson, D. (1987) *An Idea of Europe*, London: Chatto & Windus.

Holdich, T.W. (1916) *Political Frontiers and Boundary-making*, London: Macmillan.

House, J.W. (1959) 'The Franco-Italian boundary in the Alpes Maritimes', *Transactions of the Institute of British Geographers*, 26: 107–31.

Howard, M. (1965) *The Theory and Practice of War*, London: Temple Smith.

—— (1976) *War in European History*, Oxford: Oxford University Press.

—— (1986) *The Causes of Wars*, London: Temple Smith.

Hyde, C.C. (1933) 'Maps as evidence in international boundary disputes', *American Journal of International Law*, 27: 311–16.

Inalçik, H. (1973) *The Ottoman Empire*, New York: Praeger.

Inlow, E.B. (1964) 'The MacMahon Line', *Journal of Geography*, 63: 261–72.

Jackson, W.A.D. (1958) 'Whither political geography?', *Annals of Association of American Geographers*, 48: 178–83.

—— (1962) *The Russo-Chinese Borderlands*, Princeton, NJ: Van Nostrand.

—— (1971) *Politics and Geographic Relationships*, Englewood Cliffs, NJ: Prentice-Hall.

Jackson, W.G.F. (1957) *Seven Roads to Moscow*, London: Eyre & Spottiswoode.'

James, J.R., House, J.W. and Hall, P. (1970) 'Local government reform in England — symposium', *Geographical Journal*, 136: 1–23.

Johnson, R.B. (1949) 'Political salients and transportation solutions', *Annals of Association of American Geographers*, 39: 71–2.

Johnston, R.J. (1982) *Geography and the State — An Essay in Political Geography*, London: Macmillan.

Joll, J. (1969) 'Europe — a historian's view', *Montagu Burton Lecture*, University of Leeds.

Jones, S.B. (1945) *Boundary-making — A Handbook for Statesmen, Treaty Editors and Boundary Commissioners*, Washington, DC: Carnegie Endowment for International Peace.

—— (1954) 'Power inventory and national strategy', *World Politics*, 6: 421–52.

—— (1955) 'Views of the political world', *Geographical Review*, 45: 309–26.

—— (1955) 'Global strategic views', *Geographical Review*, 45: 492–508.

—— (1959) 'Boundary concepts in the setting of place and time', *Annals of Association of American Geographers*, 49: 241–55.

Kapil, R., (1966) 'On the conflict of inherited boundaries in Africa', *World Politics*, 18: 656–73.

Kasperson, R.E. and Minghi, J.V. (eds) (1969) *The Structure of Political Geography*, Chicago: Aldine.

Kedourie, E. (1966) *Nationalism*, London: Hutchinson.

Kerner, R.J. (1942) *Russia's Urge to the Sea*, Cambridge: Cambridge University Press.

Kirk, W. (1962) 'The inner Asian frontiers of India', *Transactions of the Institute of British Geographers*, 31: 131–68.

—— (1965) 'Geographical pivots of history', inaugural lecture, Leicester: Leicester University Press.

Kissinger, H. (ed.) (1968) *Problems of National Strategy*, New York: Praeger.

Kjellén, R. (1917) *Der Staat als Lebensform*, Leipzig, Hirzel.

Klineberg, O. (1964) *The Human Dimension in International Relations*, New York: Holt, Rinehart & Winston.

Kliot, N. and Waterman, S. (eds) (1983) *Pluralism and Political Geography — People, Territory and State*, London: Croom Helm.

Knight, D.B. (1982) 'Identity and territory — geographical perspectives on nationalism and regionalism', *Annals of Association of American Geographers*, 72: 514–31.

—— (1983) 'Self-determination as a geopolitical force', *Journal of Geography*, 82: 148–52.

Knorr, K. (1956) *The War Potential of Nations*, Princeton, NJ: Princeton University Press.

Kohn, H. (1945) *The Idea of Nationalism*, London: Macmillan.

—— (1953) *Pan-Slavism — Its History and Ideology*, Notre Dame, Ind.: Notre Dame University Press.

Kolinsky, M. (ed.) (1978) *Divided Loyalties*, Manchester: Manchester University Press.

Krallert, W. (1961) 'Methodische Probleme der Völker — und Sprachenkarte', *International Yearbook of Cartography*, 1: 99–120.

Krenz, F.E. (1961) *International Enclaves and Rights of Passage*, Paris: Droz et Minard.

Kristof, L.K.D. (1959) 'The nature of frontiers and boundaries', *Annals of Association of American Geographers*, 49: 269–82.

—— (1960) 'The origin and evolution of geopolitics', *Journal of Conflict Resolution*, 4: 15–51.

—— (1967) 'The state idea, the national idea and the image of the fatherland', *Orbis*, 11: 238–55.

Lapradelle, P. (1928) *La Frontière — Etude du Droit International*, Paris: Editions Internationales.

Lattimore, O. (1937) 'Origins of the Great Wall of China — a frontier concept in theory and practice', *Geographical Review*, 27: 529–49.

—— (1953) 'The new political geography of Inner Asia', *Geographical Journal*, 119: 17–32.

Laufer, H. (1985) *Das föderative System der BRD*, Munich: Bayerische Landeszentrale für Politische Bildung.

Legg, S. (1970) *The Heartland*, London: Secker & Warburg.

Leonski, Z. (1978) 'Aspects of territorial subdivisions in European socialist states', *International Social Science Journal*, 30: 47–56.

Le Page, R.B. (1964) *The National Language Question*, Oxford: Oxford University Press.

Levinson, C. (1980) *Vodka-Cola*, London: Gordon & Cremonesi.

Lichtheim, G. (1971) *Imperialism*, London: Penguin.

Liddell-Hart, B.H. (1968) *Strategy — The Indirect Approach*, New York: Praeger.

Lipman, V.D. (1949) *Local Government Areas 1834–1945*, Oxford: Blackwell.

Loth, W. (1987) *The Division of the World*, London: Croom Helm.

Luard, E. (1982) *A History of the United Nations*, vol. 1, London: HMSO.

Lubasz, H. (1964) *The Development of the Modern State*, New York: Macmillan.

Luttwak, E.N. (1974) *The Political Uses of Seapower*, Baltimore: Johns Hopkins University Press.

—— (1979) *The Grand Strategy of the Roman Empire 1 AD — 300 AD*, Baltimore: Johns Hopkins University Press.

Macartney, C.A. (1934) *National States and National Minorities*, Oxford: Oxford University Press.

McColl, R.W. (1969) 'The insurgent state-territorial bases of revolution', *Annals of Association of American Geographers*, 59: 613–31.

McElwee, W. (1974) *The Art of War — Waterloo to Mons*, London: Weidenfeld & Nicolson.

Mackenroth, G. (1953) *Bevölkerungslehre*, Berlin: Springer.

Mackinder, H.J. (1904) 'The geographical pivot of history', *Geographical Journal*, 23: 421–42.

—— (1942) *Democratic Ideals and Reality*, New York: Holt, Rinehart & Winston.

Mackintosh, J.P. (1969) *The Devolution of Power — Local Democracy, Regionalism and Nationalism*, London: Penguin.

McLennan, G., Held, D. and Hall, S. (eds) (1984) *The Idea of the Modern State*, Milton Keynes: Open University.

MacMahon, A. (ed.) (1955) *Federalism Mature and Emergent*, New York: Doubleday.

MacMahon, H. (1935) 'International boundaries', *Journal of the Royal Society of Arts*, 84: 2–16.

McNeil, W.H. (1964) *The Rise of the West*, New York: Mentor.

Mance, O. (1946) *Frontiers, Peace Treaties and International Organization*, Oxford: Oxford University Press.

Maull, O. (1928) 'Politische Grenzen', *Weltpolitische Bücherei*, 13: Berlin, Zentral-Verlag.

—— (1941) *Das Wesen der Geopolitik*, Leipzig: Teubner.

Meinig, D.W. (1956) 'Heartland and rimland in Eurasian history', *Western Political Quarterly*, 9: 553–69.

Melamid, A. (1960) 'Partitioning Cyprus — A classic example in applied geography', *Journal of Geography*, 59: 118–22.

Mellor, R.E.H. (1971) *Comecon*, Princeton, NJ: Van Nostrand.

Mikesell, M.W. (1960) 'Comparative studies in frontier history', *Annals of Association of American Geographers*, 50: 62–74.

Miksche, F.O. (1944) *Blitzkrieg — Etude sur la Tactique Allemande de 1937 à 1943*, London: Penguin.

Montgomery, B.L. (1968) *A History of Warfare*, London: Collins.

Moodie, A.E. (1945) *The Italo-Yugoslav Boundary*, London: Philip.

Moore, J. (1979) *Seapower and Politics*, London: Weidenfeld & Nicolson.

Muir, R. (1975) *Modern Political Geography*, London: Macmillan.

Nef, J.U. (1950) *War and Human Progress*, London: Routledge & Kegan Paul.

Newbigin, M.I. (1915) *Geographical Aspects of the Balkan Problems in their Relation to the Great European War*, New York: Putnam.

Niebuhr, R. (1959) *The Structure of Nations and Empires*, New York: Scribner.

—— (1981) *North Atlantic Treaty Organization — Facts and Figures*, Brussels: NATO Information Service.

Oman, C.W.C. (1924) *The Art of War in the Middle Ages*, 2 vols, London: Methuen.

O'Sullivan, P. and Miller, J. (1983) *The Geography of Warfare*, London: Croom Helm.

—— (1983) 'A geographical analysis of guerrilla warfare', *Political Geography Quarterly*, 2: 139–50.

—— (1986) *Geopolitics*, London: Croom Helm.

Paddison, R. (1983) *The Fragmented State — A Political Geography of Power*, Oxford: Blackwell.

Parker, W.H. (1982) *Mackinder — Geography as an Aid to Statecraft*, Oxford: Oxford University Press.

Pathmanathan, K.M. (1978) 'The political dynamics of ASEAN cooperation', *Indonesian Quarterly*, 6: 28–41.

Paxton, J. (ed.) *The Statesman's Yearbook*, annually, London: Macmillan.

Peake, H.J.E. (1930) 'Geographical aspects of administrative areas', *Geography*, 15: 531–46.

Pearcy, G.E. (1957) *World Political Geography*, London: Constable.

—— (1977) *World Sovereignty*. Fullerton, Ca.: Plycon.

Peltier, L.C. and Pearcy, G.E. (1966) *Military Geography*, Princeton, NJ: Van Nostrand.

Pepper, D. and Jenkins, A. (1985) *The Geography of Peace and War*, Oxford: Blackwell.

Phillips, T.R. (ed.) (1943) *Roots of Strategy*, London: Bodley Head.

Pirenne, H. (1959) *Medieval Cities*, New York: Doubleday.

Platt, D.C.M. (ed.) (1977) *Business Imperialism 1840–1930*, Oxford: Oxford University Press.

Pounds, N.J.G. (1951) 'The origin of the idea of natural frontiers in France', *Annals of Association of American Geographers*, 41: 146–57.

—— (1954) 'France and "les limites naturelles" from the seventeenth to the twentieth centuries', *Annals of Association of American Geographers*, 44: 51–62.

—— (1959) 'A free and secure access to the sea', *Annals of Association of American Geographers*, 49: 256–68.

—— (1963) *Political Geography*, New York: McGraw-Hill.

Pounds, N.J.G. and Ball, S.S. (1964) 'Core-areas and the development of the European states system', *Annals of Association of American Geographers*, 54: 24–40.

Prescott, J.R.V. (1959) 'Function and method of electoral geography', *Annals of Association of American Geographers*, 49: 296–304.

—— (1965) *The Geography of Frontiers and Boundaries*, London: Hutchinson.

—— (1969) *The Geography of State Policies*, London: Hutchinson.

—— (1972) *Political Geography*, London: Methuen.

—— (1975)*The Political Geography of the Oceans*, Newton Abbot: David & Charles.

—— (1986) *The Maritime Political Boundaries of the World*, London: Methuen.

—— (1987) *Political Frontiers and Boundaries*, London: Allen & Unwin.

Prochazka, T. (1961) 'The delimitation of the Czechoslovak-German frontier after Munich', *Journal of Central European Affairs*, 21: 100–218.

Quam, L.O. (1943) 'The use of maps in propaganda', *Journal of Geography*, 42: 21–32.

Ratzel, F. (1896) 'Die Gesetze des räumlichen Wachstums der Staaten', *Petermanns Mitteilungen*, 42: 97–107.

—— (1897) *Politische Geographie — oder die Geographie der Staaten, des Verkehrs und des Krieges*, Berlin: Oldenbourg.

Renfrew, C. (1975) 'Trade as action at a distance' in Sabloff, J.A. and Lamberg-Karlovsky, C. (eds) *Ancient Civilization and Trade*, Albuquerque, N. Mex.: University of New Mexico Press.

—— (1977) 'Space, time and polity' in Friedman, J. and Rowlands, J. (eds)

The Evolution of Social Systems, New York: Academic Press.

Rivkin, A. (1969) *Nation-building in Africa*, New Brunswick, NJ: Rutgers University Press.

Robinson, E.A.G. (ed.) (1960) *The Economic Consequences of the Size of Nations*, London: Macmillan.

Robinson, G.W.S. (1953) 'Berlin — the geography of an exclave', Geographical Review, 43: 540–57.

—— (1959) 'Exclaves', *Annals of Association of American Geographers*, 49: 283–95.

Rokkan, S. and Urwin, D. (eds) (1982) *The Politics of Territorial Identity — Studies in European Regionalism*, London: Sage.

Roskill, S.W. (1962) *The Strategy of Seapower*, London: Collins.

Rothenburg, G.E. (1977) *The Art of Warfare in the Age of Napoleon*, London: Arms & Armour Press.

Royal Commission (1969) *Local Government in England*, 2 vols, (Redcliffe-Maud), London: HMSO.

Royal Commission (1969) *Local Government in Scotland*, 2 vols, (Wheatley), Edinburgh: HMSO.

Royal Institute of International Affairs (1939) *Nationalism*, London.

Russett, B.M. (1967) *International Regions and International Systems — A Study in Political Ecology*, Chicago: Rand McNally.

Ryan, C. (1965) 'The French-Canadian dilemma', *Foreign Affairs*, 43: 462–74.

Sanquin, A.L. (1977) *La Géographie Politique*, Paris: Presses Universitaires de France.

—— (1977) 'Géographie politique — espace aérien et cosmos', *Annales de Géographie*, 86: 257–78.

—— (1983) 'Territorial aspects of federation', *Scottish Geographical Magazine*, 99: 66–75.

Schmidt, P. (1951) *Hitler's Interpreter*, London: Heinemann.

Schofer, J.P. (1975) 'Territoriality at the micro, meso- and macro-scale', *Journal of Geography*, 74: 151–8.

Schöller, P. (1957) 'Wege und Irrwege der Politischen Geographie', *Erdkunde*, 11: 1–20.

Schwind, M. (1950) *Landschaft und Grenze — Geographische Betrachtungen der deutsch-niederländischen Grenze*, Bielefeld, Verlhagen & Klasing.

—— (1970) 'Die Aufgaben einer Politischen Geographie in neuer Sicht, *Geographische Rundschau*, 22: 97–103.

—— (1972) *Allgemeine Staatengeographie*, Berlin: de Gruyter.

Servan-Schreiber, J.J. (1967) *Le Défi Americain*, Paris: Denoel.

Service, E.R. (1975) *Origins of the State and Civilization*, New York: Norton.

Seton Watson, H. (1977) *Nations and States*, London: Methuen.

Shaw, M.N. (1986) *Title to Territory in Africa — International Legal Issues*, Oxford: Oxford University Press.

Siedentop, I. (1964) 'Geographie der Enklaven und Exklaven', *Zeitschrift für Wirtschaftsgeographie*, 12: 12–14.

Siegfried, A. (1952) *The Character of Peoples*, London: Cape.

Sikorski, W. (1942) *Modern Warfare*, London: Hutchinson.

Small, M. and Singer, J.D. (1982) *Resort to Arms — International and Civil Wars 1816–1980*, London: Sage.

Smith, A.D. (1986) *The Ethnic Origin of Nations*, Oxford: Blackwell.

Smith, C.G. (1958) 'Arab nationalism — a study in political geography', *Geography*, 43: 229–42.

Smith, T. (1981) *The Pattern of Imperialism*, Cambridge: Cambridge University Press.

Soja, E.W. (1971) 'The political organisation of space', *Association of American Geographers*, Resource Paper 8, New York.

Sölch, J. (1924) *Die Auffassung der 'natürlichen Grenzen' in der wissenschaftlichen Geographie*, Innsbruck: Wagner.

Sombart, W. (1928) *Volk und Raum*, Hamburg: Hanseatische Verlagsanstalt.

Soppelsa, J. (1980) *Géographie des Armaments*, Paris: Masson.

Spengler, O. (1932) *Der Untergang des Abendlandes*, Munich: Beck.

Sprout, H.H. and Sprout, M. (eds) (1945) *Foundations of National Power*, Princeton, NJ: Princeton University Press.

—— (1966) *The Ecological Perspective on Human Affairs — with Special Reference to International Politics*, Oxford: Oxford University Press.

Spykman, N.J. (1942) 'Frontiers, security and international organization', *Geographical Review*, 32: 436–47.

—— (1970) *America's Strategy in World Politics — The United States and the Balance of Power* (originally published 1942), Hamden, Conn.: Archon.

Strachan, H. (1983) *European Armies and the Conduct of War*, London: Allen & Unwin.

Strausz-Hupé, R. (1942) *Geopolitics — The Struggle for Space and Power*, New York: Putnam.

Sukhural, B.L. (1971) *India — A Political Geography*, New Delhi: Allied Publishers.

Taylor, A.J.P. (1948) *The Habsburg Monarchy 1809–1918*, London: Hamish Hamilton.

Taylor, P.J. (1985) *Political Geography, World-Economy, Nation-State and Locality*, London: Longman.

Taylor, P.J. and House, J.W. (eds) (1984) *Political Geography — Recent Advances and Future Directions*, London: Croom Helm.

Tenbrock, R.H. (1949) *Deutsche Geschichte*, Paderborn: Schöningh.

Tilly, C. (ed.) (1975) *The Formation of Nation-States in Western Europe*, Princeton, NJ: Princeton University Press.

Tivey, L. (ed.) (1981) *The Nation-State*, Oxford: Robertson.

Toynbee, A.J. (1946, 1957) *A Study of History: Abridgement by D.C. Somervell* (vols.I-VI 1946, VII-X 1957), Oxford: Oxford University Press.

Tugendhat, C. (1973) *The Multinationals*, London: Penguin.

Turner, F.J. (1953) *The Frontier in American History* (3rd impression) New York: Holt, Rinehart & Winston.

van Houtte, H. (1935) 'Frontières naturelles et principe des nationalités', *Bulletin, Academie Royale de Belgique*, Série 5: 21.

Voyenne, B. (1964) *Histoire de l'Idée Européenne*, Paris: Payot.

Wallerstein, I. (1974) *The Modern World-System*, New York: Academic Press.

—— (1979) *The Capitalist World-Economy*, Cambridge: Cambridge University Press.

Weigert, H.W. (1942) *Generals and Geographers — The Twilight of Geopolitics*, New York: Oxford University Press.

Weigert, H.W. and Stefansson, V. (eds) (1945) *Compass of the World —
A Symposium on Political Geography*, New York: Macmillan.

Weigert, H.W. and Harrison, R.E. (eds) (1949) *New Compass of the World
— A Symposium on Political Geography*, New York: Macmillan.

Weigert, H.W. and associates (1957) *Principles of Political Geography*, New
York: Appleton-Century-Crofts.

Wheare, K.C. (1953) *The Statute of Westminster and Dominion Status*,
Oxford: Oxford University Press.

—— (1953) *Federalism*, Oxford: Oxford University Press.

Whebell, C.F.J. (1970) 'Models of political territory', *Annals of Association of American Geographers*, 60: 152–6.

Whittlesey, D. (1939) *The Earth and State — A Study in Political Geography*,
New York: Holt.

Wilkinson, H.R. (1951) *Maps and Politics — Ethnographic Cartography of
Macedonia*, Liverpool: Liverpool University Press.

Wolfe, R.I. (1963) *Transportation and Politics*, Princeton, NJ: Van Nostrand.

Index